T0185627

SpringerBriefs in Mathematics

SpringerBriefs in Mathematics showcases expositions in all areas of mathematics and applied mathematics. Manuscripts presenting new results or a single new result in a classical field, new field, or an emerging topic, applications, or bridges between new results and already published works, are encouraged. The series is intended for mathematicians and applied mathematicians. All works are peer-reviewed to meet the highest standards of scientific literature.

BCAM SpringerBriefs

Editorial Board

BCAM *SpringerBriefs* aims to publish contributions in the following disciplines: Applied Mathematics, Finance, Statistics and Computer Science. BCAM has appointed an Editorial Board, who evaluate and review proposals.

Typical topics include: a timely report of state-of-the-art analytical techniques, bridge between new research results published in journal articles and a contextual literature review, a snapshot of a hot or emerging topic, a presentation of core concepts that students must understand in order to make independent contributions.

Please submit your proposal to the Editorial Board or to Francesca Bonadei, Executive Editor Mathematics, Statistics, and Engineering: francesca.bonadei@springer.com.

More information about this series at http://www.springer.com/series/10030

Jingrui Sun · Jiongmin Yong

Stochastic Linear-Quadratic Optimal Control Theory: Differential Games and Mean-Field Problems

Jingrui Sun
Department of Mathematics
Southern University of Science
and Technology
Shenzhen, Guangdong, China

Jiongmin Yong
Department of Mathematics
University of Central Florida
Orlando, FL, USA

ISSN 2191-8198   ISSN 2191-8201   (electronic)
SpringerBriefs in Mathematics
ISBN 978-3-030-48305-0   ISBN 978-3-030-48306-7   (eBook)
https://doi.org/10.1007/978-3-030-48306-7

This Springer imprint is published by the registered company Springer Nature Switzerland AG
The registered company address is: Gewerbestrasse 11, 6330 Cham, Switzerland

To Our Parents
Yuqi Sun and Xiuying Ma
Wenyao Yong and Xiangxia Chen

Preface

Linear-quadratic optimal control theory (LQ theory, for short) has a long history, and most people feel that the LQ theory is quite mature. Three well-known relevant issues are involved: existence of optimal controls, solvability of the optimality system (which is a two-point boundary value problem), and solvability of the associated Riccati equation. A rough impression is that these three issues are somehow equivalent.

In the past few years, we, together with our collaborators, have been re-investigating the LQ theory for stochastic systems with deterministic coefficients. A number of interesting delicate issues have been identified, including:

- For finite-horizon LQ problems, open-loop optimal controls and closed-loop optimal strategies should be distinguished because the existence of the latter implies the existence of the former, but not vice versa. Whereas, for infinite-horizon LQ problems, under proper conditions, the open-loop and closed-loop solvability are equivalent.
- For finite-horizon two-person (not necessarily zero-sum) differential games, the open-loop and closed-loop Nash equilibria are two different concepts. The existence of one of them does not imply the existence of the other, which is different from LQ optimal control problems.
- The closed-loop representation of an open-loop Nash equilibrium is not necessarily the outcome of a closed-loop Nash equilibrium.

Our investigations also revealed some previously unknown facts concerning two-person differential games. A partial list is:

- For two-person (not necessarily zero-sum) differential games in finite horizons, the existence of an open-loop Nash equilibrium is equivalent to the solvability of a system of coupled FBSDEs, together with the convexities of the cost functionals; the existence of a closed-loop Nash equilibrium is equivalent to the solvability of a Lyapunov-Riccati type equation.

- For two-person zero-sum differential games, both in finite and infinite horizons, if closed-loop saddle points exist, and an open-loop saddle point exists and admits a closed-loop representation, then the representation must be the outcome of some closed-loop saddle point. Such a result also holds for LQ optimal control problems.
- For two-person zero-sum differential games over an infinite horizon, the existence of an open-loop and a closed-loop Nash equilibrium are equivalent.
- Some of the results concerning LQ optimal control problems can further be extended to the case when expectations of the state and the control are involved. This kind of LQ problems is referred to as the mean-field problem.

The purpose of this book is to systematically present the above-mentioned results concerning LQ differential games and mean-field LQ optimal control problems. We assume that readers are familiar with basic stochastic analysis and stochastic control theory.

This work is supported in part by NSFC Grant 11901280 and NSF Grants DMS-1406776, DMS-1812921.

The authors also would like to express their gratitude to the anonymous referees for their constructive comments, which led to this improved version.

Shenzhen, China   Jingrui Sun
Orlando, USA   Jiongmin Yong
March 2020

Contents

Frequently Used Notation

I. Notation for Euclidean Spaces and Matrices

1. $\mathbb{R}^{n \times m}$: the space of all $n \times m$ real matrices.
2. $\mathbb{R}^n = \mathbb{R}^{n \times 1}$; $\mathbb{R} = \mathbb{R}^1$; $\overline{\mathbb{R}} = [-\infty, \infty]$.
3. $\mathbb{S}^n$: the space of all symmetric $n \times n$ real matrices.
4. $\mathbb{S}^n_+$: the subset of $\mathbb{S}^n$ consisting of positive definite matrices.
5. $\overline{\mathbb{S}}^n_+$: the subset of $\mathbb{S}^n$ consisting of positive semi-definite matrices.
6. I_n: the identity matrix of size n, which is also denoted simply by I if no confusion occurs.
7. $M^\top$: the transpose of a matrix M.
8. $M^\dagger$: the Moore-Penrose pseudoinverse of a matrix M.
9. $\text{tr}(M)$: the sum of diagonal elements of a square matrix M, called the trace of M.
10. $\langle \cdot, \cdot \rangle$: the inner product on a Hilbert space. In particular, the usual inner product on $\mathbb{R}^{n \times m}$ is given by $\langle M, N \rangle \mapsto \text{tr}(M^\top N)$.
11. $|M| \triangleq \sqrt{\text{tr}(M^\top M)}$: the Frobenius norm of a matrix M.
12. $\mathscr{R}(M)$: the range of a matrix or an operator M.
13. $\mathscr{N}(M)$: the kernel of a matrix or an operator M.
14. $A \geqslant B$: $A - B$ is a positive semi-definite symmetric matrix.
15. $\mathcal{Q}(P) \triangleq PA + A^\top P + C^\top PC + Q$.
16. $\mathcal{S}(P) \triangleq B^\top P + D^\top PC + S$.
17. $\mathcal{R}(P) \triangleq R + D^\top PD$.
18. $\widehat{A} \triangleq A + \bar{A}$, $\widehat{B} \triangleq B + \bar{B}$, $\widehat{C} \triangleq C + \bar{C}$, $\widehat{D} \triangleq D + \bar{D}$.
19. $\widehat{Q} \triangleq Q + \bar{Q}$, $\widehat{S} \triangleq S + \bar{S}$, $\widehat{R} \triangleq R + \bar{R}$, $\widehat{G} \triangleq G + \bar{G}$.
20. $\widehat{\mathcal{Q}}(P, \Pi) \triangleq \Pi\widehat{A} + \widehat{A}^\top \Pi + \widehat{C}^\top P\widehat{C} + \widehat{Q}$.
21. $\widehat{\mathcal{S}}(P, \Pi) \triangleq \widehat{B}^\top \Pi + \widehat{D}^\top P\widehat{C} + \widehat{S}$.
22. $\widehat{\mathcal{R}}(P) \triangleq \widehat{R} + \widehat{D}^\top P\widehat{D}$.

II. Sets and Spaces of Functions and Processes

Let $\mathbb{H}$ be a Euclidian space (which could be $\mathbb{R}^n$, $\mathbb{R}^{n \times m}$, etc.).

1. $C([t, T]; \mathbb{H})$: the space of $\mathbb{H}$-valued, continuous functions on $[t, T]$.
2. $L^p(t, T; \mathbb{H})$: the space of $\mathbb{H}$-valued functions that are pth ($1 \leqslant p < \infty$) power Lebesgue integrable on $[t, T]$.
3. $L^\infty(t, T; \mathbb{H})$: the space of $\mathbb{H}$-valued, Lebesgue measurable functions that are essentially bounded on $[t, T]$.
4. $L^2_{\mathcal{F}_t}(\Omega; \mathbb{H})$: the space of $\mathcal{F}_t$-measurable, $\mathbb{H}$-valued random variables ξ such that $\mathbb{E}|\xi|^2 < \infty$.
5. $L^2_{\mathbb{F}}(\Omega; L^1(t, T; \mathbb{H}))$: the space of $\mathbb{F}$-progressively measurable, $\mathbb{H}$-valued processes $\varphi : [t, T] \times \Omega \to \mathbb{H}$ such that $\mathbb{E}\left[\int_t^T |\varphi(s)| ds\right]^2 < \infty$.
6. $L^2_{\mathbb{F}}(t, T; \mathbb{H})$: the space of $\mathbb{F}$-progressively measurable, $\mathbb{H}$-valued processes $\varphi : [t, T] \times \Omega \to \mathbb{H}$ such that $\mathbb{E}\int_t^T |\varphi(s)|^2 ds < \infty$.
7. $L^2_{\mathbb{F}}(\mathbb{H})$: the space of $\mathbb{F}$-progressively measurable, $\mathbb{H}$-valued processes $\varphi : [0, \infty) \times \Omega \to \mathbb{H}$ such that $\mathbb{E}\int_0^\infty |\varphi(t)|^2 dt < \infty$.
8. $L^2_{\mathbb{F}}(\Omega; C([t, T]; \mathbb{H}))$: the space of $\mathbb{F}$-adapted, continuous, $\mathbb{H}$-valued processes $\varphi : [t, T] \times \Omega \to \mathbb{H}$ such that $\mathbb{E}\left[\sup_{s \in [t, T]} |\varphi(s)|^2\right] < \infty$.
9. $\mathcal{X}_t = L^2_{\mathcal{F}_t}(\Omega; \mathbb{R}^n)$.
10. $\mathcal{X}[t, T] = L^2_{\mathbb{F}}(\Omega; C([t, T]; \mathbb{R}^n))$.
11. $\mathcal{U}[t, T] = L^2_{\mathbb{F}}(t, T; \mathbb{R}^m)$.
12. $\mathcal{X}_{loc}[0, \infty) = \bigcap_{T > 0} \mathcal{X}[0, T]$.
13. $\mathcal{X}[0, \infty)$: the subspace of $\mathcal{X}_{loc}[0, \infty)$ consisting of processes φ which are square-integrable: $\mathbb{E}\int_0^\infty |\varphi(t)|^2 dt < \infty$.

Chapter 1
Some Elements of Linear-Quadratic Optimal Controls

.

Abstract This chapter is a brief review on the stochastic linear-quadratic optimal control. Some useful concepts and results, which will be needed throughout this book, are presented in the context of finite and infinite horizon problems. These materials are mainly for beginners and may also serve as a quick reference for knowledgeable readers.

Keywords Linear-quadratic · Optimal control · Finite horizon · Infinite horizon · Riccati equation · Open-loop · Closed-loop

In this chapter, we briefly review the stochastic linear-quadratic (LQ, for short) optimal control problem and present some useful concepts and results in the context of finite and infinite horizon problems. Most of the results recalled here are quoted from the book [48] by Sun and Yong, where rigorous proofs can be found. In the sequel, $(\Omega, \mathcal{F}, \mathbb{P})$ denotes a complete probability space on which a standard one-dimensional Brownian motion $W = \{W(t); 0 \leqslant t < \infty\}$ is defined, and $\mathbb{F}$ denotes the usual augmentation of the natural filtration $\{\mathcal{F}_t\}_{t \geqslant 0}$ generated by W. For a random variable ξ, we write $\xi \in \mathcal{F}_t$ if ξ is $\mathcal{F}_t$-measurable, and for a stochastic process φ, we write $\varphi \in \mathbb{F}$ if it is $\mathbb{F}$-progressively measurable.

1.1 LQ Optimal Control Problems in Finite Horizons

Consider the following controlled linear stochastic differential equation (SDE, for short) on a finite horizon $[t, T]$:

$$\begin{cases} dX(s) = [A(s)X(s) + B(s)u(s) + b(s)]ds \\ \qquad\qquad + [C(s)X(s) + D(s)u(s) + \sigma(s)]dW(s), \qquad (1.1.1) \\ X(t) = x, \end{cases}$$

where $A, C : [0, T] \to \mathbb{R}^{n \times n}$, $B, D : [0, T] \to \mathbb{R}^{n \times m}$, called the *coefficients* of the *state equation* (1.1.1), and $b, \sigma : [0, T] \times \Omega \to \mathbb{R}^n$, called the *nonhomogeneous terms*, satisfy the following assumption.

© The Author(s), under exclusive license to Springer Nature Switzerland AG 2020
J. Sun and J. Yong, *Stochastic Linear-Quadratic Optimal Control Theory:*
Differential Games and Mean-Field Problems, SpringerBriefs in Mathematics,
https://doi.org/10.1007/978-3-030-48306-7_1

(H1) The coefficients and the nonhomogeneous terms of (1.1.1) satisfy

$$\begin{cases} A \in L^1(0, T; \mathbb{R}^{n\times n}), & B \in L^2(0, T; \mathbb{R}^{n\times m}), & b \in L^2_{\mathbb{F}}(\Omega; L^1(0, T; \mathbb{R}^n)), \\ C \in L^2(0, T; \mathbb{R}^{n\times n}), & D \in L^\infty(0, T; \mathbb{R}^{n\times m}), & \sigma \in L^2_{\mathbb{F}}(0, T; \mathbb{R}^n). \end{cases}$$

In the above assumption we have adopted the following notation: For a subset $\mathbb{H}$ of some Euclidean space (which could be $\mathbb{R}^n$, $\mathbb{R}^{n\times m}$, etc.),

$$L^p(t, T; \mathbb{H}) = \left\{ \varphi : [t, T] \to \mathbb{H} \;\Big|\; \int_t^T |\varphi(s)|^p ds < \infty \right\} \quad (1 \leqslant p < \infty),$$

$$L^\infty(t, T; \mathbb{H}) = \left\{ \varphi : [t, T] \to \mathbb{H} \;\Big|\; \varphi \text{ is essentially bounded} \right\},$$

$$L^2_{\mathbb{F}}(t, T; \mathbb{H}) = \left\{ \varphi : [t, T] \times \Omega \to \mathbb{H} \;\Big|\; \varphi \in \mathbb{F}, \mathbb{E}\int_t^T |\varphi(s)|^2 ds < \infty \right\},$$

$$L^2_{\mathbb{F}}(\Omega; L^1(t, T; \mathbb{H})) = \left\{ \varphi : [t, T] \times \Omega \to \mathbb{H} \;\Big|\; \varphi \in \mathbb{F}, \mathbb{E}\Big[\int_t^T |\varphi(s)| ds\Big]^2 < \infty \right\}.$$

The process $u = \{u(s); t \leqslant s \leqslant T\}$ in (1.1.1) belongs to the space

$$\mathcal{U}[t, T] \equiv L^2_{\mathbb{F}}(t, T; \mathbb{R}^m)$$

and is called a *control*.

The *cost functional* associated with the state equation (1.1.1) is of the quadratic form

$$\begin{aligned} J(t, x; u) = \mathbb{E}\bigg\{ &\langle GX(T), X(T)\rangle + 2\langle g, X(T)\rangle \\ &+ \int_t^T \bigg[\bigg\langle \begin{pmatrix} Q(s) & S(s)^\top \\ S(s) & R(s) \end{pmatrix} \begin{pmatrix} X(s) \\ u(s) \end{pmatrix}, \begin{pmatrix} X(s) \\ u(s) \end{pmatrix} \bigg\rangle \\ &+ 2\bigg\langle \begin{pmatrix} q(s) \\ \rho(s) \end{pmatrix}, \begin{pmatrix} X(s) \\ u(s) \end{pmatrix} \bigg\rangle\bigg] ds \bigg\}. \end{aligned} \tag{1.1.2}$$

In the above, $\langle \cdot, \cdot \rangle$ stands for the Frobenius inner product of two matrices that have the same size. That is, for $M, N \in \mathbb{R}^{n\times m}$, $\langle M, N\rangle$ is equal to the trace of $M^\top N$. The superscript $\top$ denotes the transpose of matrices. In the sequel, the identity matrix of size n will be denoted by I_n, and the Frobenius norm of a matrix M will be denoted by $|M|$. Let $\mathbb{S}^n$ (respectively, $\mathbb{S}^n_+$) be the space of all symmetric $n \times n$ real matrices (respectively, positive definite matrices), and let

$$L^2_{\mathcal{F}_t}(\Omega; \mathbb{R}^n) = \left\{ \xi : \Omega \to \mathbb{R}^n \mid \xi \text{ is } \mathcal{F}_t\text{-measurable with } \mathbb{E}|\xi|^2 < \infty \right\}.$$

The *weighting matrices* in the cost functional are assumed to satisfy the following condition.

(H2) The weighting matrices in the cost functional (1.1.2) satisfy

$$\begin{cases} Q \in L^1(0, T; \mathbb{S}^n), & S \in L^2(0, T; \mathbb{R}^{m \times n}), \\ R \in L^\infty(0, T; \mathbb{S}^m), & q \in L^2_{\mathbb{F}}(\Omega; L^1(0, T; \mathbb{R}^n)), \\ \rho \in L^2_{\mathbb{F}}(0, T; \mathbb{R}^m), & g \in L^2_{\mathcal{F}_T}(\Omega; \mathbb{R}^n), \quad G \in \mathbb{S}^n. \end{cases}$$

Under (H1) and (H2), for any *initial pair* $(t, x) \in [0, T) \times \mathbb{R}^n$ and control $u \in \mathcal{U}[t, T]$, the solution X of the state equation (1.1.1) belongs to the space

$$L^2_{\mathbb{F}}(\Omega; C([t, T]; \mathbb{R}^n)) = \Big\{ \varphi : [t, T] \times \Omega \to \mathbb{R}^n \mid \varphi \text{ is } \mathbb{F}\text{-adapted, continuous,}$$

$$\text{and } \mathbb{E}\Big[\sup_{t \leqslant s \leqslant T} |X(s)|^2 \Big] < \infty \Big\},$$

and hence the integral on the right-hand side of (1.1.2) is well-defined. The *stochastic LQ optimal control problem* (SLQ problem, for short) is to minimize $J(t, x; u)$ subject to the state equation (1.1.1). A precise formulation of the SLQ problem is as follows.

Problem (SLQ). For a given initial pair $(t, x) \in [0, T) \times \mathbb{R}^n$, find a control $u^* \in \mathcal{U}[t, T]$ such that

$$J(t, x; u^*) = \inf_{u \in \mathcal{U}[t, T]} J(t, x; u). \tag{1.1.3}$$

Definition 1.1.1 A process $u^* \in \mathcal{U}[t, T]$ such that (1.1.3) holds is called an *open-loop optimal control* for the initial pair (t, x). When such a u^* exists, we say Problem (SLQ) is *open-loop solvable at* (t, x).

Definition 1.1.2 Problem (SLQ) is said to be (*uniquely*) *open-loop solvable* if for every initial pair $(t, x) \in [0, T) \times \mathbb{R}^n$, a (unique) open-loop optimal control exists.

Definition 1.1.3 The function $V : [0, T] \times \mathbb{R}^n \to \mathbb{R}$ defined by

$$V(t, x) \triangleq \inf_{u \in \mathcal{U}[t, T]} J(t, x; u)$$

is called the *value function* of Problem (SLQ). Problem (SLQ) is said to be *finite* if $V(t, x) > -\infty$ for every initial pair (t, x).

Remark 1.1.4 It is clear that the process u^* in (1.1.3) depends on the initial pari (t, x). The notion of an open-loop optimal control is employed to emphasize this dependence.

Write for simplicity $\boldsymbol{\Theta}[t, T] = L^2(t, T; \mathbb{R}^{m \times n})$. Let $(\Theta, v) \in \boldsymbol{\Theta}[t, T] \times \mathcal{U}[t, T]$, and let X be the solution to the SDE

$$
\begin{cases}
dX(s) = [(A + B\Theta)X + Bv + b]ds \\
\qquad\quad + [(C + D\Theta)X + Dv + \sigma]dW, \quad s \in [t, T], \\
X(t) = x,
\end{cases}
\tag{1.1.4}
$$

where, for notational convenience, we have suppressed the argument s in the drift and diffusion terms. The process

$$
u(s) = \Theta(s)X(s) + v(s), \quad s \in [t, T]
\tag{1.1.5}
$$

is easily seen to be square-integrable and hence is a control. Plugging (1.1.5) into the SDE (1.1.4), we see that (1.1.4) is exactly the form (1.1.1).

Definition 1.1.5 A *closed-loop strategy* over $[t, T]$ is a pair $(\Theta, v) \in \boldsymbol{\Theta}[t, T] \times \mathcal{U}[t, T]$. The process defined by (1.1.5) is called a *closed-loop control* generated by (Θ, v), and the SDE (1.1.4) is called a *closed-loop system*.

Remark 1.1.6 A closed-loop strategy (Θ, v) is independent of the initial state x. However, the closed-loop control generated by (Θ, v) still depends on x.

Definition 1.1.7 A closed-loop strategy $(\Theta, v) \in \boldsymbol{\Theta}[t, T] \times \mathcal{U}[t, T]$ is said to be *optimal* if for every initial state, the closed-loop control generated by (Θ, v) is open-loop optimal. In this case, (Θ, v) is called a *closed-loop optimal strategy*.

Definition 1.1.8 Problem (SLQ) is said to be (*uniquely*) *closed-loop solvable* over $[t, T]$, if a (unique) closed-loop optimal strategy exists on $[t, T]$.

In order to study Problem (SLQ), it is often convenient to consider first the *homogeneous problem* associated with Problem (SLQ), for which we want to find a control $u \in \mathcal{U}[t, T]$ such that

$$
J^0(t, x; u) \triangleq \mathbb{E}\left\{ \langle GX(T), X(T) \rangle + \int_t^T \left\langle \begin{pmatrix} Q & S^\top \\ S & R \end{pmatrix} \begin{pmatrix} X \\ u \end{pmatrix}, \begin{pmatrix} X \\ u \end{pmatrix} \right\rangle ds \right\},
$$

is minimized subject to

$$
\begin{cases}
dX(s) = [A(s)X(s) + B(s)u(s)]ds \\
\qquad\quad + [C(s)X(s) + D(s)u(s)]dW(s), \quad s \in [t, T], \\
X(t) = x.
\end{cases}
$$

We denote this homogeneous problem by Problem $(SLQ)^0$ and its value function by $V^0(t, x)$.

Now we present a characterization of open-loop optimal controls.

Theorem 1.1.9 *Let* (H1)–(H2) *hold, and let* $(t, x) \in [0, T) \times \mathbb{R}^n$ *be a given initial pair. Then* $u \in \mathcal{U}[t, T]$ *is an open-loop optimal control of Problem (SLQ) for* (t, x) *if and only if*

(i) *the mapping* $v \mapsto J^0(t, 0; v)$ *is convex, or equivalently,*

$$J^0(t, 0; v) \geqslant 0, \quad \forall v \in \mathcal{U}[t, T]; \tag{1.1.6}$$

(ii) *the adapted solution* (X, Y, Z) *to the decoupled forward-backward stochastic differential equation (FBSDE, for short)*

$$\begin{cases} dX(s) = (AX + Bu + b)ds + (CX + Du + \sigma)dW, \\ dY(s) = -\big(A^\top Y + C^\top Z + QX + S^\top u + q\big)ds + ZdW, \\ X(t) = x, \quad Y(T) = GX(T) + g \end{cases}$$

satisfies the stationarity condition

$$B^\top Y + D^\top Z + SX + Ru + \rho = 0, \quad \text{a.e. on } [t, T], \text{ a.s.}$$

The closed-loop optimal strategy can be characterized by means of the *Riccati equation*, which is a nonlinear ordinary differential equation (ODE, for short) of the following form:

$$\begin{cases} \dot{P}(s) + \mathcal{Q}(s, P(s)) - \mathcal{S}(s, P(s))^\top \mathcal{R}(s, P(s))^\dagger \mathcal{S}(s, P(s)) = 0, \\ P(T) = G, \end{cases} \tag{1.1.7}$$

where the superscript † denotes the *Moore–Penrose pseudoinverse* of matrices, and

$$\begin{cases} \mathcal{Q}(s, P) \triangleq PA(s) + A(s)^\top P + C(s)^\top PC(s) + Q(s), \\ \mathcal{S}(s, P) \triangleq B(s)^\top P + D(s)^\top PC(s) + S(s), \\ \mathcal{R}(s, P) \triangleq R(s) + D(s)^\top PD(s). \end{cases} \tag{1.1.8}$$

The Eq. (1.1.7) is symmetric, so by a solution we mean a continuous $\mathbb{S}^n$-valued function that satisfies (1.1.7) for almost all s. To simplify notation we will frequently suppress the variable s and write $\mathcal{Q}(s, P(s))$, $\mathcal{S}(s, P(s))$, and $\mathcal{R}(s, P(s))$ as $\mathcal{Q}(P)$, $\mathcal{S}(P)$, and $\mathcal{R}(P)$, respectively.

Let $\mathscr{R}(M)$ denotes the range of a matrix M, and let $C([t, T]; \mathbb{S}^n)$ be the space of all $\mathbb{S}^n$-valued, continuous functions over $[t, T]$.

Definition 1.1.10 Let $P \in C([t, T]; \mathbb{S}^n)$ be a solution to the Riccati equation (1.1.7) on $[t, T]$. We say that the solution P is *regular* if

(i) $\mathcal{R}(P)^\dagger \mathcal{S}(P) \in \Theta[t, T]$;
(ii) $\mathscr{R}(\mathcal{S}(P)) \subseteq \mathscr{R}(\mathcal{R}(P))$ a.e. on $[t, T]$;
(iii) $\mathcal{R}(P) \geqslant 0$ a.e. on $[t, T]$, i.e., $\mathcal{R}(P)$ is positive semi-definite a.e. on $[t, T]$.

The Riccati equation (1.1.7) is said to be *regularly solvable* when a regular solution exists.

Definition 1.1.11 A solution P of the Riccati equation (1.1.7) on $[t, T]$ is said to be *strongly regular* if there exists a constant $\lambda > 0$ such that

$$R(s) + D(s)^\top P(s)D(s) \geqslant \lambda I_m, \quad \text{a.e. } s \in [t, T].$$

The Riccati equation (1.1.7) is said to be *strongly regularly solvable* when a strongly regular solution exists.

The following result provides a characterization of closed-loop optimal strategies.

Theorem 1.1.12 *Let* (H1)–(H2) *hold. Then Problem* (SLQ) *is closed-loop solvable on* $[t, T]$ *if and only if*

(i) *the Riccati equation* (1.1.7) *admits a regular solution* $P \in C([t, T]; \mathbb{S}^n)$;
(ii) *with*
$$\Theta(s) \triangleq -\mathcal{R}(s, P(s))^\dagger \mathcal{S}(s, P(s)), \quad s \in [t, T],$$

the adapted solution (η, ζ) *to the backward stochastic differential equation* (BSDE, for short)

$$\begin{cases} d\eta(s) = -\big[(A + B\Theta)^\top \eta + (C + D\Theta)^\top \zeta + (C + D\Theta)^\top P\sigma \\ \qquad\qquad + \Theta^\top \rho + Pb + q\big]ds + \zeta dW, \quad s \in [t, T], \\ \eta(T) = g \end{cases} \tag{1.1.9}$$

satisfies the following conditions:

$$\kappa \triangleq B^\top \eta + D^\top \zeta + D^\top P\sigma + \rho \in \mathscr{R}(\mathcal{R}(P)), \quad \text{a.e. on } [t, T], \text{ a.s.,}$$
$$v \triangleq -\mathcal{R}(P)^\dagger \kappa \in \mathcal{U}[t, T].$$

In this case, the closed-loop optimal strategy (Θ^*, v^*) *admits the following representation:*

$$\Theta^* = \Theta + [I - \mathcal{R}(P)^\dagger \mathcal{R}(P)]\Pi, \quad v^* = v + [I - \mathcal{R}(P)^\dagger \mathcal{R}(P)]\pi,$$

where $(\Pi, \pi) \in \mathbf{\Theta}[t, T] \times \mathcal{U}[t, T]$ *is arbitrary. Moreover,*

$$V(t, x) = \mathbb{E}\Big\{ \langle P(t)x, x \rangle + 2\langle \eta(t), x \rangle + \int_t^T \Big[\langle P(s)\sigma(s), \sigma(s) \rangle + 2\langle \eta(s), b(s) \rangle$$
$$+ 2\langle \zeta(s), \sigma(s) \rangle - \langle \mathcal{R}(s, P(s))^\dagger \kappa(s), \kappa(s) \rangle \Big]ds \Big\}.$$

An equivalent statement of Theorem 1.1.12 is as follows.

Theorem 1.1.13 *Let* (H1)–(H2) *hold. Then a closed-loop strategy* $(\Theta, v) \in$ $\Theta[t, T] \times \mathcal{U}[t, T]$ *is optimal if and only if*

(i) *the solution* $P \in C([t, T]; \mathbb{S}^n)$ *to the symmetric Lyapunov type equation*

$$\begin{cases} \dot{P} + \mathcal{Q}(P) + \Theta^\top \mathcal{R}(P)\Theta + \mathcal{S}(P)^\top \Theta + \Theta^\top \mathcal{S}(P) = 0, \\ P(T) = G, \end{cases}$$

satisfies the following two conditions: for almost all $s \in [t, T]$,

$$\mathcal{R}(P) \geqslant 0, \quad \mathcal{S}(P) + \mathcal{R}(P)\Theta = 0;$$

(ii) *the adapted solution* (η, ζ) *to the BSDE*

$$\begin{cases} d\eta(s) = -\big[(A + B\Theta)^\top \eta + (C + D\Theta)^\top \zeta + (C + D\Theta)^\top P\sigma \\ \qquad\quad + \Theta^\top \rho + Pb + q\big]ds + \zeta dW(s), \quad s \in [t, T], \\ \eta(T) = g, \end{cases}$$

satisfies the following condition: for almost all $s \in [t, T]$,

$$B^\top \eta + D^\top \zeta + D^\top P\sigma + \rho + \mathcal{R}(P)v = 0, \quad \text{a.s.}$$

The closed-loop solvability trivially implies the open-loop solvability of Problem (SLQ). An example has been constructed in [48] to show that the converse implication does not hold in general. The following result gives a condition under which these two types of solvability are equivalent.

Theorem 1.1.14 *Let* (H1)–(H2) *hold. Suppose that there exists a constant* $\lambda > 0$ *such that*

$$J^0(0, 0; u) \geqslant \lambda \mathbb{E} \int_0^T |u(s)|^2 ds, \quad \forall u \in \mathcal{U}[0, T]. \tag{1.1.10}$$

Then the following hold:

(i) *Problem* (SLQ) *is uniquely open-loop solvable.*
(ii) *The Riccati equation* (1.1.7) *admits a unique strongly regular solution* $P \in$ $C([0, T]; \mathbb{S}^n)$ *and hence Problem* (SLQ) *is uniquely closed-loop solvable over* $[t, T]$ *for every* $t \in [0, T)$.
(iii) *The unique closed-loop optimal strategy over each* $[t, T]$ *is given by the following unified form:*

$$\Theta^* = -\mathcal{R}(P)^{-1}\mathcal{S}(P),$$
$$v^* = -\mathcal{R}(P)^{-1}(B^\top \eta + D^\top \zeta + D^\top P\sigma + \rho),$$

where (η, ζ) *is the adapted solution of* (1.1.9) *with* Θ *replaced by* Θ^*.

(iv) The unique open-loop optimal control for the initial pair (t, x) is the closed-loop control generated by the closed-loop optimal strategy (Θ^*, v^*) over $[t, T]$.

It should be noted that part of the converse of Theorem 1.1.14 is true. In fact, we have the following result.

Theorem 1.1.15 *Let* (H1)–(H2) *hold. Then the following are equivalent:*

(i)   there exists a constant $\lambda > 0$ such that (1.1.10) *holds;*
(ii) the Riccati equation (1.1.7) *admits a strongly regular solution on $[0, T]$;*
(iii) there exists an $\mathbb{S}^n$-valued function P such that

$$\mathcal{R}(t, P(t)) \geqslant \lambda I_m, \quad a.e. \ t \in [0, T]$$

holds for some constant $\lambda > 0$ and

$$V^0(t, x) = \langle P(t)x, x \rangle, \quad \forall (t, x) \in [0, T] \times \mathbb{R}^n.$$

1.2   LQ Optimal Control Problems in Infinite Horizons

Let $L_{\mathbb{F}}^2(\mathbb{R}^k)$ be the space of $\mathbb{R}^k$-valued, $\mathbb{F}$-progressively measurable processes that are square-integrable on $[0, \infty)$. Consider the controlled linear SDE

$$\begin{cases} dX(t) = [AX(t) + Bu(t) + b(t)]dt + [CX(t) + Du(t) + \sigma(t)]dW, \\ X(0) = x, \end{cases}$$

over the infinite horizon $[0, \infty)$ and the quadratic cost functional

$$J(x; u) \triangleq \mathbb{E} \int_0^\infty \left[\left\langle \begin{pmatrix} Q & S^\top \\ S & R \end{pmatrix} \begin{pmatrix} X(t) \\ u(t) \end{pmatrix}, \begin{pmatrix} X(t) \\ u(t) \end{pmatrix} \right\rangle + 2 \left\langle \begin{pmatrix} q(t) \\ \rho(t) \end{pmatrix}, \begin{pmatrix} X(t) \\ u(t) \end{pmatrix} \right\rangle \right] dt,$$

where

$$A, C \in \mathbb{R}^{n \times n}, \quad B, D \in \mathbb{R}^{n \times m}, \quad Q \in \mathbb{S}^n, \quad S \in \mathbb{R}^{m \times n}, \quad R \in \mathbb{S}^m$$

are given constant matrices, and

$$b, \sigma, q \in L_{\mathbb{F}}^2(\mathbb{R}^n), \quad \rho \in L_{\mathbb{F}}^2(\mathbb{R}^m)$$

are given processes. In the above, the control process u belongs to the space $L_{\mathbb{F}}^2(\mathbb{R}^m)$. A control u is said to be *admissible* with respect to the initial state x if the corresponding state process $\{X(t; x, u); 0 \leqslant t < \infty\}$ satisfies

$$\mathbb{E}\int_0^\infty |X(t;x,u)|^2 dt < \infty.$$

The set of admissible controls with respect to x is denoted by $\mathcal{U}_{ad}(x)$.

The linear-quadratic optimal control problem over $[0, \infty)$ can be stated as follows.

Problem $(SLQ)_\infty$. For a given initial state $x \in \mathbb{R}^n$, find an admissible control $u^* \in \mathcal{U}_{ad}(x)$ such that

$$J(x; u^*) = \inf_{u \in \mathcal{U}_{ad}(x)} J(x; u) \equiv V(x).$$

In the special case of $b, \sigma, q, \rho = 0$, we denote Problem $(SLQ)_\infty$ by Problem $(SLQ)_\infty^0$, the cost functional by $J^0(x; u)$, and the value function by $V^0(x)$.

In order to obtain conditions under which the admissible control sets $\mathcal{U}_{ad}(x)$ are nonempty, we introduce the following concepts.

Definition 1.2.1 The uncontrolled system

$$dX(t) = AX(t)dt + CX(t)dW(t), \quad t \geqslant 0 \tag{1.2.1}$$

is said to be L^2-*stable* if for every initial state $x \in \mathbb{R}^n$, its solution $X(t; x)$ is square-integrable on $[0, \infty)$, that is,

$$\mathbb{E}\int_0^\infty |X(t;x)|^2 dt < \infty, \quad \forall x \in \mathbb{R}^n.$$

Definition 1.2.2 The controlled system

$$dX(t) = [AX(t) + Bu(t)]dt + [CX(t) + Du(t)]dW(t), \quad t \geqslant 0 \tag{1.2.2}$$

is said to be L^2-*stabilizable* if there exists a matrix $\Theta \in \mathbb{R}^{m \times n}$ such that the uncontrolled system

$$dX(t) = (A + B\Theta)X(t)dt + (C + D\Theta)X(t)dW(t), \quad t \geqslant 0$$

is L^2-stable. In this case, Θ is called a *stabilizer* of (1.2.2). The set of all stabilizers of (1.2.2) is denoted by $\mathscr{S} \equiv \mathscr{S}[A, C; B, D]$.

Theorem 1.2.3 *The following statements are equivalent:*

(i) $\mathcal{U}_{ad}(x) \neq \varnothing$ *for all* $x \in \mathbb{R}^n$;
(ii) $\mathscr{S}[A, C; B, D] \neq \varnothing$;
(iii) *The following algebraic Riccati equation (ARE, for short) admits a positive definite solution* $P \in \mathbb{S}_+^n$:

$$PA + A^\top P + C^\top PC + I$$
$$- (PB + C^\top PD)(I + D^\top PD)^{-1}(B^\top P + D^\top PC) = 0. \qquad (1.2.3)$$

If the above are satisfied and $P \in \mathbb{S}_+^n$ is a solution of (1.2.3), *then*

$$\Gamma \triangleq -(I + D^\top PD)^{-1}(B^\top P + D^\top PC) \in \mathscr{S}[A, C; B, D].$$

According to Theorem 1.2.3, Problem (SLQ)$_\infty$ is well-posed (for all $x \in \mathbb{R}^n$) only if (1.2.2) is L^2-stabilizable. So it is reasonable to assume the following:

(S) The system (1.2.2) is L^2-stabilizable, i.e., $\mathscr{S}[A, C; B, D] \neq \varnothing$.

Definition 1.2.4 An element $u^* \in \mathcal{U}_{ad}(x)$ is called an *open-loop optimal control* of Problem (SLQ)$_\infty$ for the initial state $x \in \mathbb{R}^n$ if

$$J(x; u^*) \leqslant J(x; u), \quad \forall u \in \mathcal{U}_{ad}(x).$$

If an open-loop optimal control (uniquely) exists for x, Problem (SLQ)$_\infty$ is said to be (*uniquely*) *open-loop solvable at x*. Problem (SLQ)$_\infty$ is said to be (*uniquely*) *open-loop solvable* if it is (uniquely) open-loop solvable at all $x \in \mathbb{R}^n$.

Definition 1.2.5 A pair $(\Theta, v) \in \mathscr{S}[A, C; B, D] \times L_{\mathbb{F}}^2(\mathbb{R}^m)$ is called a *closed-loop strategy* of Problem (SLQ)$_\infty$. The outcome

$$u(t) \triangleq \Theta X(t) + v(t), \quad t \geqslant 0$$

of a closed-loop strategy (Θ, v) is called a *closed-loop control* for the initial state x.

Definition 1.2.6 A closed-loop strategy (Θ^*, v^*) is said to be *optimal* if

$$J(x; \Theta^* X^* + v^*) \leqslant J(x; \Theta X + v),$$

for all $(x, \Theta, v) \in \mathbb{R}^n \times \mathscr{S}[A, C; B, D] \times L_{\mathbb{F}}^2(\mathbb{R}^m)$. If a closed-loop optimal strategy (uniquely) exists, Problem (SLQ)$_\infty$ is said to be (*uniquely*) *closed-loop solvable*.

Now similar to (1.1.8), we define for a constant matrix $P \in \mathbb{S}^n$ that

$$\begin{cases} \mathcal{Q}(P) \triangleq PA + A^\top P + C^\top PC + Q, \\ \mathcal{S}(P) \triangleq B^\top P + D^\top PC + S, \\ \mathcal{R}(P) \triangleq R + D^\top PD. \end{cases} \qquad (1.2.4)$$

Note that in the above, all matrices are time-invariant. In order to study the open-loop and closed-loop solvability of Problem (SLQ)$_\infty$, we further introduce the following concept.

Definition 1.2.7 The following constrained nonlinear algebraic equation

$$\begin{cases} \mathcal{Q}(P) - \mathcal{S}(P)^\top \mathcal{R}(P)^\dagger \mathcal{S}(P) = 0, \\ \mathscr{R}(\mathcal{S}(P)) \subseteq \mathscr{R}(\mathcal{R}(P)), \\ \mathcal{R}(P) \geqslant 0, \end{cases} \tag{1.2.5}$$

with the unknown $P \in \mathbb{S}^n$, is called a *generalized algebraic Riccati equation* (GARE, for short). A solution P of (1.2.5) is said to be *stabilizing* if there exists a $\Pi \in \mathbb{R}^{m \times n}$ such that the matrix

$$\Theta \triangleq -\mathcal{R}(P)^\dagger \mathcal{S}(P) + [I - \mathcal{R}(P)^\dagger \mathcal{R}(P)]\Pi$$

is a stabilizer of the system (1.2.2).

The main result on the solvability of Problem $(SLQ)_\infty$ is as follows.

Theorem 1.2.8 *Let (S) hold. Then the following statements are equivalent:*

(i) *Problem $(SLQ)_\infty$ is open-loop solvable;*
(ii) *Problem $(SLQ)_\infty$ is closed-loop solvable;*
(iii) *The GARE (1.2.5) admits a stabilizing solution $P \in \mathbb{S}^n$, and the BSDE*

$$\begin{aligned} d\eta(t) = -\Big\{ &[A - B\mathcal{R}(P)^\dagger \mathcal{S}(P)]^\top \eta + [C - D\mathcal{R}(P)^\dagger \mathcal{S}(P)]^\top \zeta \\ &+ [C - D\mathcal{R}(P)^\dagger \mathcal{S}(P)]^\top P\sigma - \mathcal{S}(P)^\top \mathcal{R}(P)^\dagger \rho \\ &+ Pb + q \Big\} dt + \zeta dW, \quad t \geqslant 0 \end{aligned}$$

admits an L^2-stable adapted solution[1] (η, ζ) such that

$$\theta(t) \triangleq B^\top \eta(t) + D^\top \zeta(t) + D^\top P\sigma(t) + \rho(t) \in \mathscr{R}(\mathcal{R}(P)),$$
$$\text{a.e. } t \in [0, \infty), \text{ a.s.}$$

In the above case, all closed-loop optimal strategies (Θ^*, v^*) are given by

$$\Theta^* = -\mathcal{R}(P)^\dagger \mathcal{S}(P) + [I - \mathcal{R}(P)^\dagger \mathcal{R}(P)]\Pi,$$
$$v^* = -\mathcal{R}(P)^\dagger \theta + [I - \mathcal{R}(P)^\dagger \mathcal{R}(P)]\nu,$$

where $\Pi \in \mathbb{R}^{m \times n}$ is chosen so that $\Theta^* \in \mathscr{S}[A, C; B, D]$ and $\nu \in L_\mathbb{F}^2(\mathbb{R}^m)$ is arbitrary; every open-loop optimal control u^* for the initial state x admits a closed-loop representation:

$$u^*(t) = \Theta^* X^*(t) + v^*(t), \quad t \geqslant 0,$$

---

[1]See the next section for the notion of an L^2-stable adapted solution to BSDEs over an infinite horizon.

where (Θ^*, v^*) is a closed-loop optimal strategy of Problem (SLQ)$_\infty$ *and X^* is the corresponding closed-loop state process. Moreover,*

$$V(x) = \langle Px, x \rangle + 2\mathbb{E}\langle \eta(0), x \rangle$$
$$+ \mathbb{E} \int_0^\infty \left[\langle P\sigma, \sigma \rangle + 2\langle \eta, b \rangle + 2\langle \zeta, \sigma \rangle - \langle \mathcal{R}(P)^\dagger \theta, \theta \rangle \right] dt.$$

1.3 Appendix: Pseudoinverse and Infinite-Horizon BSDEs

Let M be an $m \times n$ real matrix. The *Moore-Penrose pseudoinverse* of M, denoted by $M^\dagger$, is an $n \times m$ real matrix such that

$$MM^\dagger M = M, \quad (MM^\dagger)^\top = MM^\dagger,$$
$$M^\dagger MM^\dagger = M^\dagger, \quad (M^\dagger M)^\top = M^\dagger M.$$

Every matrix has a unique (Moore-Penrose) pseudoinverse. If $M \in \mathbb{S}^n$, then $M^\dagger \in \mathbb{S}^n$, $MM^\dagger = M^\dagger M$, and $M \geqslant 0$ if and only if $M^\dagger \geqslant 0$.

Proposition 1.3.1 *Let $\mathcal{I}$ be an interval. Let $L(t)$ and $N(t)$ be two Lebesgue measurable functions on $\mathcal{I}$, with values in $\mathbb{R}^{n\times k}$ and $\mathbb{R}^{n\times m}$, respectively. Then the equation $N(t)X(t) = L(t)$ has a solution $X(t) \in L^2(\mathcal{I}; \mathbb{R}^{m\times k})$ if and only if*

$$(i) \ \mathscr{R}(L(t)) \subseteq \mathscr{R}(N(t)), \quad and \quad (ii) \ N(t)^\dagger L(t) \in L^2(\mathcal{I}; \mathbb{R}^{m\times k}),$$

in which case the general solution is given by

$$X(t) = N(t)^\dagger L(t) + [I_m - N(t)^\dagger N(t)]Y(t),$$

where $Y(t) \in L^2(\mathcal{I}; \mathbb{R}^{m\times k})$ is arbitrary.

Remark 1.3.2 The following are obvious:

(i) The condition $\mathscr{R}(L(t)) \subseteq \mathscr{R}(N(t))$ is equivalent to

$$N(t)N(t)^\dagger L(t) = L(t).$$

(ii) If $N(t) \in \mathbb{S}^n$ and $N(t)X(t) = L(t)$, then

$$X(t)^\top N(t)X(t) = L(t)^\top N(t)^\dagger L(t).$$

Next, let $\mathcal{X}[0, \infty)$ be the subspace of $L^2_{\mathbb{F}}(\mathbb{R}^n)$ whose elements are continuous. Consider the following BSDE over the infinite horizon $[0, \infty)$:

$$dY(t) = -[A^\top Y(t) + C^\top Z(t) + \varphi(t)]dt + Z(t)dW(t), \quad t \in [0, \infty), \quad (1.3.1)$$

where $A, C \in \mathbb{R}^{n \times n}$ are given constant matrices, and $\{\varphi(t); 0 \leqslant t < \infty\}$ is a given $\mathbb{F}$-progressively measurable, $\mathbb{R}^n$-valued process.

Definition 1.3.3 An L^2-*stable adapted solution* to the BSDE (1.3.1) is a pair $(Y, Z) \in \mathcal{X}[0, \infty) \times L^2_{\mathbb{F}}(\mathbb{R}^n)$ that satisfies the integral version of (1.3.1):

$$Y(t) = Y(0) - \int_0^t \left[A^\top Y(s) + C^\top Z(s) + \varphi(s) \right] ds + \int_0^t Z(s)dW(s), \ t \geqslant 0.$$

The following theorem establishes the basic existence and uniqueness result for the BSDE (1.3.1).

Theorem 1.3.4 *Suppose that the system* (1.2.1) *is* L^2-*stable. Then for every* $\varphi \in L^2_{\mathbb{F}}(\mathbb{R}^n)$, *the BSDE* (1.3.1) *admits a unique* L^2-*stable adapted solution.*

Chapter 2
Linear-Quadratic Two-Person Differential Games

Abstract The purpose of this chapter is to develop a theory for stochastic linear-quadratic two-person differential games. Open-loop and closed-loop Nash equilibria are explored in the context of nonzero-sum and zero-sum differential games. The existence of an open-loop Nash equilibrium is characterized in terms of a system of constrained forward-backward stochastic differential equations, and the existence of a closed-loop Nash equilibrium is characterized by the solvability of a system of coupled symmetric differential Riccati equations. It is shown that in the nonzero-sum case, the closed-loop representation for open-loop Nash equilibria is different from the outcome of closed-loop Nash equilibria in general, whereas they coincide in the zero-sum case when both exist. Some results for infinite-horizon zero-sum differential games are also established in terms of algebraic Riccati equation.

Keywords Linear-quadratic · Differential game · Two-person · Nonzero-sum · Zero-sum · Nash equilibrium · Saddle point · Open-loop · Closed-loop · Riccati equation

Throughout this chapter, $(\Omega, \mathcal{F}, \mathbb{P})$ is a complete probability space in the background, and $\mathbb{F} = \{\mathcal{F}_t\}_{t \geqslant 0}$ is a filtration over $[0, \infty)$. As usual, we assume that the probability space $(\Omega, \mathcal{F}, \mathbb{P})$ is rich enough to support a standard one-dimensional Brownian motion $W = \{W(t); 0 \leqslant t < \infty\}$, and that $\mathbb{F}$ is the usual augmentation of the natural filtration generated by W. We employ throughout this chapter the notation of Chap. 1.

2.1 Formulation

Consider the controlled linear SDE on a finite time horizon $[t, T]$:

$$\begin{cases} dX(s) = [A(s)X(s) + B_1(s)u_1(s) + B_2(s)u_2(s) + b(s)]ds \\ \qquad\quad + [C(s)X(s) + D_1(s)u_1(s) + D_2(s)u_2(s) + \sigma(s)]dW(s), \quad (2.1.1) \\ X(t) = x, \end{cases}$$

where A, B_1, B_2, C, D_1, D_2, b and σ are given coefficients. We suppose that in this chapter, the following assumption holds.

(G1) The coefficients and the nonhomogeneous terms of (2.1.1) satisfy

$$\begin{cases} A \in L^1(0, T; \mathbb{R}^{n \times n}), & B_i \in L^2(0, T; \mathbb{R}^{n \times m_i}), & i = 1, 2, \\ C \in L^2(0, T; \mathbb{R}^{n \times n}), & D_i \in L^\infty(0, T; \mathbb{R}^{n \times m_i}), & i = 1, 2, \\ b \in L^2_{\mathbb{F}}(\Omega; L^1(0, T; \mathbb{R}^n)), & \sigma \in L^2_{\mathbb{F}}(0, T; \mathbb{R}^n). \end{cases}$$

In the above, X is the n-dimensional state process, and u_i ($i = 1, 2$), valued in $\mathbb{R}^{m_i}$, is the control process selected by Player i. For $i = 1, 2$ and $t \in [0, T)$, the set of *admissible (open-loop) controls* of Player i on $[t, T]$ is defined by

$$\mathcal{U}_i[t, T] = \left\{ u_i : [t, T] \times \Omega \to \mathbb{R}^{m_i} \mid u_i \in \mathbb{F} \text{ and } \mathbb{E} \int_t^T |u_i(s)|^2 ds < \infty \right\},$$

and the cost functional for Player i is defined by

$$\begin{aligned} J^i(t, x; u_1, u_2) = \mathbb{E} \Bigg\{ & \langle G^i X(T), X(T) \rangle + 2 \langle g^i, X(T) \rangle \\ &+ \int_t^T \Bigg[\left\langle \begin{pmatrix} Q^i(s) & S_1^i(s)^\top & S_2^i(s)^\top \\ S_1^i(s) & R_{11}^i(s) & R_{12}^i(s) \\ S_2^i(s) & R_{21}^i(s) & R_{22}^i(s) \end{pmatrix} \begin{pmatrix} X(s) \\ u_1(s) \\ u_2(s) \end{pmatrix}, \begin{pmatrix} X(s) \\ u_1(s) \\ u_2(s) \end{pmatrix} \right\rangle \\ &+ 2 \left\langle \begin{pmatrix} q^i(s) \\ \rho_1^i(s) \\ \rho_2^i(s) \end{pmatrix}, \begin{pmatrix} X(s) \\ u_1(s) \\ u_2(s) \end{pmatrix} \right\rangle \Bigg] ds \Bigg\}. \end{aligned} \qquad (2.1.2)$$

We assume that the weighting coefficients satisfy the following assumption so that the integral on the right-hand side makes sense:

(G2) The weighting coefficients satisfy the following conditions:

$$\begin{cases} G^i \in \mathbb{S}^n, & g^i \in L^2_{\mathcal{F}_T}(\Omega; \mathbb{R}^n), & Q^i \in L^1(0, T; \mathbb{S}^n), \\ S_j^i \in L^2(0, T; \mathbb{R}^{m_j \times n}), & R_{jj}^i \in L^\infty(0, T; \mathbb{S}^{m_j}), \\ q^i \in L^2_{\mathbb{F}}(\Omega; L^1(0, T; \mathbb{R}^n)), & \rho_j^i \in L^2_{\mathbb{F}}(0, T; \mathbb{R}^{m_j}), \\ R_{12}^i = (R_{21}^i)^\top \in L^\infty(0, T; \mathbb{R}^{m_1 \times m_2}); & i, j = 1, 2. \end{cases}$$

The problem we are concerned with in this chapter is the following linear-quadratic stochastic *two-person differential game*.

Problem (SDG). For given initial pair $(t, x) \in [0, T) \times \mathbb{R}^n$, how the players should choose their controls $u_i \in \mathcal{U}_i[t, T]$ so that their payoff $J^i(t, x; u_1, u_2)$ is minimized.

The above problem is called an LQ stochastic two-person *zero-sum* differential game if one player's gain is the other's loss, or equivalently, the sum of the winnings and losses of the players is always zero, i.e.,

$$J^1(t, x; u_1, u_2) + J^2(t, x; u_1, u_2) = 0, \quad \forall u_i \in \mathcal{U}_i[t, T], \; i = 1, 2. \qquad (2.1.3)$$

If (2.1.3) is not necessarily true, we shall call Problem (SDG) an LQ stochastic two-person *nonzero-sum* differential game, emphasizing that (2.1.3) is not assumed.

Remark 2.1.1 The LQ optimal control problem studied in Chap. 1, Sect. 1.1 can be regarded as a special case of Problem (SDG) where $m_1 = 0$ or $m_2 = 0$.

For notational simplicity, we let $m = m_1 + m_2$ and denote

$$B = (B_1, B_2), \qquad D = (D_1, D_2), \qquad R_1^i = (R_{11}^i, R_{12}^i), \qquad R_2^i = (R_{21}^i, R_{22}^i),$$
$$S^i = \begin{pmatrix} S_1^i \\ S_2^i \end{pmatrix}, \qquad R^i = \begin{pmatrix} R_{11}^i & R_{12}^i \\ R_{21}^i & R_{22}^i \end{pmatrix}, \qquad \rho^i = \begin{pmatrix} \rho_1^i \\ \rho_2^i \end{pmatrix}, \qquad u = \begin{pmatrix} u_1 \\ u_2 \end{pmatrix}.$$

Naturally, we identify $\mathcal{U}[t, T] = \mathcal{U}_1[t, T] \times \mathcal{U}_2[t, T]$. With such notation, the state equation (2.1.1) can be rewritten as

$$\begin{cases} dX(s) = [A(s)X(s) + B(s)u(s) + b(s)]ds \\ \qquad\qquad + [C(s)X(s) + D(s)u(s) + \sigma(s)]dW(s), \quad s \in [t, T], \\ X(t) = x, \end{cases}$$

and the cost functionals (2.1.2) can be rewritten as

$$J^i(t, x; u) = \mathbb{E}\Bigg\{ \langle G^i X(T), X(T) \rangle + 2\langle g^i, X(T) \rangle$$
$$+ \int_t^T \Bigg[\Big\langle \begin{pmatrix} Q^i(s) & S^i(s)^\top \\ S^i(s) & R^i(s) \end{pmatrix} \begin{pmatrix} X(s) \\ u(s) \end{pmatrix}, \begin{pmatrix} X(s) \\ u(s) \end{pmatrix} \Big\rangle$$
$$+ 2\Big\langle \begin{pmatrix} q^i(s) \\ \rho^i(s) \end{pmatrix}, \begin{pmatrix} X(s) \\ u(s) \end{pmatrix} \Big\rangle \Bigg] ds \Bigg\}.$$

Definition 2.1.2 A pair $(u_1^*, u_2^*) \in \mathcal{U}_1[t, T] \times \mathcal{U}_2[t, T]$ is called an *open-loop Nash equilibrium* of Problem (SDG) for the initial pair (t, x) if

$$\begin{aligned} J^1(t, x; u_1^*, u_2^*) &\leqslant J^1(t, x; u_1, u_2^*), \quad \forall u_1 \in \mathcal{U}_1[t, T], \\ J^2(t, x; u_1^*, u_2^*) &\leqslant J^2(t, x; u_1^*, u_2), \quad \forall u_2 \in \mathcal{U}_2[t, T]. \end{aligned} \qquad (2.1.4)$$

In the zero-sum case, (u_1^*, u_2^*) is called an *open-loop saddle point*.

For $i = 1, 2$, let us denote $\boldsymbol{\Theta}_i[t, T] = L^2(t, T; \mathbb{R}^{m_i \times n})$ and write $\boldsymbol{\Theta}[t, T] = \boldsymbol{\Theta}_1[t, T] \times \boldsymbol{\Theta}_2[t, T]$. In a similar fashion to the LQ optimal control problem, we can consider, for any $\Theta = (\Theta_1^\top, \Theta_2^\top)^\top \in \boldsymbol{\Theta}_1[t, T] \times \boldsymbol{\Theta}_2[t, T]$ and $v = (v_1^\top, v_2^\top)^\top \in \mathcal{U}_1[t, T] \times \mathcal{U}_2[t, T]$, the state equation

$$
\begin{cases}
dX(s) = [(A + B\Theta)X + Bv + b]ds \\
\qquad\qquad + [(C + D\Theta)X + Dv + \sigma]dW(s), \quad s \in [t, T], \qquad (2.1.5)\\
X(t) = x,
\end{cases}
$$

and the cost functionals ($i = 1, 2$)

$$
J^i(t, x; \Theta_1 X + v_1, \Theta_2 X + v_2) = \mathbb{E}\Bigg\{ \langle G^i X(T), X(T) \rangle + 2\langle g^i, X(T) \rangle
$$
$$
+ \int_t^T \left[\left\langle \begin{pmatrix} Q^i & (S^i)^\top \\ S^i & R^i \end{pmatrix} \begin{pmatrix} X \\ \Theta X + v \end{pmatrix}, \begin{pmatrix} X \\ \Theta X + v \end{pmatrix} \right\rangle + 2\left\langle \begin{pmatrix} q^i \\ \rho^i \end{pmatrix}, \begin{pmatrix} X \\ \Theta X + v \end{pmatrix} \right\rangle \right] ds \Bigg\}.
$$

We shall call (Θ_i, v_i) a *closed-loop strategy* of Player i, and (2.1.5) the *closed-loop system* of the original system under closed-loop strategies (Θ_1, v_1) and (Θ_2, v_2) of Players 1 and 2. To emphasize that the solution X to (2.1.5) depends on (Θ_i, v_i) ($i = 1, 2$), as well as on the initial pair (t, x), we frequently write

$$
X(\cdot) = X(\cdot\,; t, x, \Theta_1, v_1, \Theta_2, v_2).
$$

The control pair (u_1, u_2) defined by

$$
u_1 = \Theta_1 X + v_1, \quad u_2 = \Theta_2 X + v_2
$$

is called the outcome of the closed-loop strategy $(\Theta_1, v_1; \Theta_2, v_2)$.

Definition 2.1.3 A 4-tuple $(\Theta_1^*, v_1^*; \Theta_2^*, v_2^*) \in \boldsymbol{\Theta}_1[t, T] \times \mathcal{U}_1[t, T] \times \boldsymbol{\Theta}_2[t, T] \times \mathcal{U}_2[t, T]$ is called a *closed-loop Nash equilibrium* of Problem (SDG) on $[t, T]$, if for any $x \in \mathbb{R}^n$ and $(\Theta_1, v_1; \Theta_2, v_2) \in \boldsymbol{\Theta}_1[t, T] \times \mathcal{U}_1[t, T] \times \boldsymbol{\Theta}_2[t, T] \times \mathcal{U}_2[t, T]$,

$$
J^1(t, x; \Theta_1^* X^* + v_1^*, \Theta_2^* X^* + v_2^*) \leqslant J^1(t, x; \Theta_1 X + v_1, \Theta_2^* X + v_2^*), \qquad (2.1.6)
$$
$$
J^2(t, x; \Theta_1^* X^* + v_1^*, \Theta_2^* X^* + v_2^*) \leqslant J^2(t, x; \Theta_1^* X + v_1^*, \Theta_2 X + v_2). \qquad (2.1.7)
$$

In the zero-sum case, $(\Theta_1^*, v_1^*; \Theta_2^*, v_2^*)$ is called a *closed-loop saddle point*.

Remark 2.1.4 Similar to the closed-loop optimal strategy of Problem (SLQ), the closed-loop Nash equilibrium is also required to be independent of the initial state x. One should note that in both (2.1.6) and (2.1.7), $X^*(\cdot) = X(\cdot\,; t, x, \Theta_1^*, v_1^*, \Theta_2^*, v_2^*)$, whereas, in (2.1.6) $X(\cdot) = X(\cdot\,; t, x, \Theta_1, v_1, \Theta_2^*, v_2^*)$, and in (2.1.7) $X(\cdot) = X(\cdot\,; t, x, \Theta_1^*, v_1^*, \Theta_2, v_2)$. Thus, the X appeared in (2.1.6) and (2.1.7) are different in general.

The following result provides some equivalent definitions of closed-loop Nash equilibrium, whose proof is similar to the case of LQ optimal control problems; see Proposition 2.1.5 of [48, Chap. 2].

Proposition 2.1.5 *Let (G1)–(G2) hold, and let* $(\Theta_1^*, v_1^*; \Theta_2^*, v_2^*) \in \boldsymbol{\Theta}_1[t, T] \times \mathcal{U}_1[t, T] \times \boldsymbol{\Theta}_2[t, T] \times \mathcal{U}_2[t, T]$. *The following are equivalent:*

(i) $(\Theta_1^*, v_1^*; \Theta_2^*, v_2^*)$ *is a closed-loop Nash equilibrium on* $[t, T]$;
(ii) for any $x \in \mathbb{R}^n$ *and* $(v_1, v_2) \in \mathcal{U}_1[t, T] \times \mathcal{U}_2[t, T]$,

$$J^1(t, x; \Theta_1^* X^* + v_1^*, \Theta_2^* X^* + v_2^*) \leqslant J^1(t, x; \Theta_1^* X + v_1, \Theta_2^* X + v_2^*),$$
$$J^2(t, x; \Theta_1^* X^* + v_1^*, \Theta_2^* X^* + v_2^*) \leqslant J^2(t, x; \Theta_1^* X + v_1^*, \Theta_2^* X + v_2);$$

(iii) for any $x \in \mathbb{R}^n$ *and* $(u_1, u_2) \in \mathcal{U}_1[t, T] \times \mathcal{U}_2[t, T]$,

$$J^1(t, x; \Theta_1^* X^* + v_1^*, \Theta_2^* X^* + v_2^*) \leqslant J^1(t, x; u_1, \Theta_2^* X + v_2^*), \qquad (2.1.8)$$
$$J^2(t, x; \Theta_1^* X^* + v_1^*, \Theta_2^* X^* + v_2^*) \leqslant J^2(t, x; \Theta_1^* X + v_1^*, u_2). \qquad (2.1.9)$$

Suppose that $(\Theta_1^*, v_1^*; \Theta_2^*, v_2^*)$ is a closed-loop Nash equilibrium of Problem (SDG) on $[t, T]$. If we denote by (u_1^*, u_2^*) the outcome of this closed-loop Nash equilibrium, i.e.,

$$u_i^* = \Theta_i^* X^* + v_i^*, \quad i = 1, 2,$$

then (2.1.8) and (2.1.9) respectively become

$$J^1(t, x; u_1^*, u_2^*) \leqslant J^1(t, x; u_1, \Theta_2^* X + v_2^*), \qquad (2.1.10)$$
$$J^2(t, x; u_1^*, u_2^*) \leqslant J^2(t, x; \Theta_1^* X + v_1^*, u_2). \qquad (2.1.11)$$

Since in (2.1.10), X corresponds to u_1 and (Θ_2^*, v_2^*), one might not have

$$u_2^* = \Theta_2^* X + v_2^*.$$

Likewise, in (2.1.11) one might not have $u_1^* = \Theta_1^* X + v_1^*$ either. Hence, comparing this with (2.1.4), we see that, in general, the outcome of a closed-loop Nash equilibrium is not necessarily an open-loop Nash equilibrium for the initial pair (t, x).

2.2 Open-Loop Nash Equilibria and Their Closed-Loop Representations

To begin our study of open-loop Nash equilibria, we observe that the open-loop controls selected by the players are free to choose from $\mathcal{U}_i[t, T]$. This makes it possible to treat the two-person differential game as two related optimal control problems. To elaborate on the idea, let us suppose that (u_1^*, u_2^*) is an open-loop Nash equilibrium of Problem (SDG) for the initial pair (t, x). Consider the following two LQ optimal control problems:

(1) To minimize
$$J(t, x; u_1) \triangleq J^1(t, x; u_1, u_2^*)$$

subject to the state equation

$$\begin{cases} dX(s) = [A(s)X(s) + B_1(s)u_1(s) + B_2(s)u_2^*(s) + b(s)]ds \\ \qquad\quad + [C(s)X(s) + D_1(s)u_1(s) + D_2(s)u_2^*(s) + \sigma(s)]dW(s), \\ X(t) = x. \end{cases}$$

(2) To minimize
$$J(t, x; u_2) \triangleq J^2(t, x; u_1^*, u_2)$$

subject to the state equation

$$\begin{cases} dX(s) = [A(s)X(s) + B_2(s)u_2(s) + B_1(s)u_1^*(s) + b(s)]ds \\ \qquad\quad + [C(s)X(s) + D_2(s)u_2(s) + D_1(s)u_1^*(s) + \sigma(s)]dW(s), \\ X(t) = x. \end{cases}$$

According to Definition 2.1.2, u_1^* is an open-loop optimal control of Problem (1), and u_2^* is an open-loop optimal control of problem (2). Thus, we may apply Theorem 1.1.9 of Chap. 1 to the above control problems. This leads to the following result.

Theorem 2.2.1 *Let* (G1)–(G2) *hold. Then* $(u_1^*, u_2^*) \in \mathcal{U}_1[t, T] \times \mathcal{U}_2[t, T]$ *is an open-loop Nash equilibrium of Problem* (SDG) *for the initial pair* (t, x) *if and only if for* $i = 1, 2,$

(a) *the adapted solution* (X^*, Y_i^*, Z_i^*) *to the FBSDE on* $[t, T]$

$$\begin{cases} dX^*(s) = (AX^* + B_1u_1^* + B_2u_2^* + b)ds \\ \qquad\qquad + (CX^* + D_1u_1^* + D_2u_2^* + \sigma)dW(s), \\ dY_i^*(s) = -[A^\top Y_i^* + C^\top Z_i^* + Q^i X^* \\ \qquad\qquad + (S_1^i)^\top u_1^* + (S_2^i)^\top u_2^* + q^i]ds + Z_i^* dW(s), \\ X^*(t) = x, \quad Y_i^*(T) = G^i X^*(T) + g^i, \end{cases} \tag{2.2.1}$$

satisfies the stationarity condition

$$B_i^\top Y_i^* + D_i^\top Z_i^* + S_i^i X^* + R_{i1}^i u_1^* + R_{i2}^i u_2^* + \rho_i^i = 0, \tag{2.2.2}$$
$$\text{a.e. } s \in [t, T], \text{ a.s.}$$

(b) *for any* $u_i \in \mathcal{U}_i[t, T],$

$$\mathbb{E}\left\{\langle G^i X_i(T), X_i(T)\rangle + \int_t^T \left\langle \begin{pmatrix} Q^i & (S_i^i)^\top \\ S_i^i & R_{ii}^i \end{pmatrix}\begin{pmatrix} X_i \\ u_i \end{pmatrix}, \begin{pmatrix} X_i \\ u_i \end{pmatrix}\right\rangle ds\right\} \geqslant 0,$$

where X_i is the solution to the SDE

$$\begin{cases} dX_i(s) = (AX_i + B_i u_i)ds + (CX_i + D_i u_i)dW(s), & s \in [t, T], \\ X_i(t) = 0, \end{cases}$$

(or equivalently, the mapping $u_i \mapsto J^i(t, x; u_1, u_2)$ is convex).

Remark 2.2.2 (1) Note that for $i = 1, 2$ the FBSDEs (2.2.1) are coupled through the relation (2.2.2). In fact, from (2.2.2) we see that

$$\begin{pmatrix} R_{11}^1 & R_{12}^1 \\ R_{21}^2 & R_{22}^2 \end{pmatrix}\begin{pmatrix} u_1^* \\ u_2^* \end{pmatrix} = -\begin{pmatrix} B_1^\top Y_1^* + D_1^\top Z_1^* + S_1^1 X^* + \rho_1^1 \\ B_2^\top Y_2^* + D_2^\top Z_2^* + S_2^2 X^* + \rho_2^2 \end{pmatrix}.$$

Thus, say, in the case that the coefficient matrix on the left-hand side is invertible, one has

$$\begin{pmatrix} u_1^* \\ u_2^* \end{pmatrix} = -\begin{pmatrix} R_{11}^1 & R_{12}^1 \\ R_{21}^2 & R_{22}^2 \end{pmatrix}^{-1}\begin{pmatrix} B_1^\top Y_1^* + D_1^\top Z_1^* + S_1^1 X^* + \rho_1^1 \\ B_2^\top Y_2^* + D_2^\top Z_2^* + S_2^2 X^* + \rho_2^2 \end{pmatrix}.$$

Plugging the above into (2.2.1), we see the coupling between these two FBSDEs.

(2) An easily verifiable condition for (b) of Theorem 2.2.1 is

$$G^i \geqslant 0, \ R_{ii}^i(s) > 0, \ Q^i(s) - S_i^i(s)^\top R_{ii}^i(s)^{-1} S_i^i(s) \geqslant 0; \quad \text{a.e. } s \in [t, T].$$

This can be seen by completing the square: For any $x \in \mathbb{R}^n$ and $u \in \mathbb{R}^{m_i}$,

$$\begin{aligned} &\langle Q^i(s)x, x\rangle + 2\langle S_i^i(s)x, u\rangle + \langle R_{ii}^i(s)u, u\rangle \\ &= \langle [Q^i(s) - S_i^i(s)^\top R_{ii}^i(s)^{-1} S_i^i(s)]x, x\rangle \\ &\quad + \langle R_{ii}^i(s)[u + R_{ii}^i(s)^{-1} S_i^i(s)x], [u + R_{ii}^i(s)^{-1} S_i^i(s)x]\rangle \\ &\geqslant 0, \quad \text{a.e. } s \in [t, T]. \end{aligned}$$

Note that for $i = 1, 2$, the conditions (b) of Theorem 2.2.1 are independent, and both can be regarded as the convexity condition for certain Problem (SLQ) with appropriate state equation and cost functional.

We now rewrite (2.2.1) and (2.2.2) in more compact forms. Recall the notation introduced right after Remark 2.1.1. Let I_k denote the identity matrix of size k, and let

$$A = \begin{pmatrix} A & 0 \\ 0 & A \end{pmatrix}, \quad B = \begin{pmatrix} B & 0 \\ 0 & B \end{pmatrix}, \quad C = \begin{pmatrix} C & 0 \\ 0 & C \end{pmatrix}, \quad D = \begin{pmatrix} D & 0 \\ 0 & D \end{pmatrix},$$

$$G = \begin{pmatrix} G^1 & 0 \\ 0 & G^2 \end{pmatrix}, \quad Q = \begin{pmatrix} Q^1 & 0 \\ 0 & Q^2 \end{pmatrix}, \quad S = \begin{pmatrix} S^1 & 0 \\ 0 & S^2 \end{pmatrix}, \quad R = \begin{pmatrix} R^1 & 0 \\ 0 & R^2 \end{pmatrix},$$

$$g = \begin{pmatrix} g^1 \\ g^2 \end{pmatrix}, \quad q = \begin{pmatrix} q^1 \\ q^2 \end{pmatrix}, \quad \rho = \begin{pmatrix} \rho^1 \\ \rho^2 \end{pmatrix}, \quad I_k = \begin{pmatrix} I_k \\ I_k \end{pmatrix}.$$

Note that the matrices G, Q, and R are symmetric. If we denote by J the $2m \times m$ matrix

$$\begin{pmatrix} I_{m_1} & 0 \\ 0 & 0 \\ 0 & 0 \\ 0 & I_{m_2} \end{pmatrix} = \begin{pmatrix} I_{m_1} & 0_{m_1 \times m_2} \\ 0_{m_2 \times m_1} & 0_{m_2 \times m_2} \\ 0_{m_1 \times m_1} & 0_{m_1 \times m_2} \\ 0_{m_2 \times m_1} & I_{m_2} \end{pmatrix},$$

then

$$BJ = \begin{pmatrix} B_1 & B_2 & 0 & 0 \\ 0 & 0 & B_1 & B_2 \end{pmatrix} \begin{pmatrix} I_{m_1} & 0 \\ 0 & 0 \\ 0 & 0 \\ 0 & I_{m_2} \end{pmatrix} = \begin{pmatrix} B_1 & 0 \\ 0 & B_2 \end{pmatrix},$$

$$DJ = \begin{pmatrix} D_1 & D_2 & 0 & 0 \\ 0 & 0 & D_1 & D_2 \end{pmatrix} \begin{pmatrix} I_{m_1} & 0 \\ 0 & 0 \\ 0 & 0 \\ 0 & I_{m_2} \end{pmatrix} = \begin{pmatrix} D_1 & 0 \\ 0 & D_2 \end{pmatrix},$$

$$J^\top S = \begin{pmatrix} I_{m_1} & 0 & 0 & 0 \\ 0 & 0 & 0 & I_{m_2} \end{pmatrix} \begin{pmatrix} S_1^1 & 0 \\ S_2^1 & 0 \\ 0 & S_1^2 \\ 0 & S_2^2 \end{pmatrix} = \begin{pmatrix} S_1^1 & 0 \\ 0 & S_2^2 \end{pmatrix},$$

$$J^\top R = \begin{pmatrix} I_{m_1} & 0 & 0 & 0 \\ 0 & 0 & 0 & I_{m_2} \end{pmatrix} \begin{pmatrix} R_1^1 & 0 \\ R_2^1 & 0 \\ 0 & R_1^2 \\ 0 & R_2^2 \end{pmatrix} = \begin{pmatrix} R_1^1 & 0 \\ 0 & R_2^2 \end{pmatrix},$$

$$J^\top \rho = \begin{pmatrix} I_{m_1} & 0 & 0 & 0 \\ 0 & 0 & 0 & I_{m_2} \end{pmatrix} \begin{pmatrix} \rho_1^1 \\ \rho_2^1 \\ \rho_1^2 \\ \rho_2^2 \end{pmatrix} = \begin{pmatrix} \rho_1^1 \\ \rho_2^2 \end{pmatrix}.$$

With the above notation and with

$$Y(s) \triangleq \begin{pmatrix} Y_1(s) \\ Y_2(s) \end{pmatrix}, \quad Z(s) \triangleq \begin{pmatrix} Z_1(s) \\ Z_2(s) \end{pmatrix},$$

we can express the FBSDEs (2.2.1) and the stationarity conditions (2.2.2) ($i = 1, 2$), respectively, more compactly as (dropping $*$)

$$\begin{cases} dX(s) = (AX + Bu + b)ds + (CX + Du + \sigma)dW(s), \\ dY(s) = -\big(A^\top Y + C^\top Z + QI_nX + S^\top I_m u + q\big)ds + ZdW(s), \quad (2.2.3) \\ X(t) = x, \quad Y(T) = GI_nX(T) + g, \end{cases}$$

and

$$J^\top\big(B^\top Y + D^\top Z + SI_nX + RI_m u + \rho\big) = 0, \quad \text{a.e. } s \in [t, T], \text{ a.s.} \quad (2.2.4)$$

From Theorem 2.2.1 we see that the open-loop Nash equilibria are determined by (2.2.3)–(2.2.4). To solve for an open-loop Nash equilibrium (u_1, u_2) from (2.2.3)–(2.2.4), we introduce the ansatz that the adapted solution (X, Y, Z) to the FBSDE (2.2.3) has the form

$$Y = \begin{pmatrix} Y_1 \\ Y_2 \end{pmatrix} = \begin{pmatrix} \Pi_1 X + \eta_1 \\ \Pi_2 X + \eta_2 \end{pmatrix} \equiv \Pi X + \eta; \quad \Pi \triangleq \begin{pmatrix} \Pi_1 \\ \Pi_2 \end{pmatrix}, \quad \eta \triangleq \begin{pmatrix} \eta_1 \\ \eta_2 \end{pmatrix},$$

where $\Pi_i : [t, T] \to \mathbb{R}^{n \times n}$; $i = 1, 2$, are differentiable maps to be determined, and $\eta : [t, T] \times \Omega \to \mathbb{R}^{2n}$ is a stochastic process satisfying a certain equation. To match the terminal condition $Y(T) = GI_nX(T) + g$, we impose the requirements

$$\Pi(T) = GI_n, \quad \eta(T) = g.$$

The second requirement suggests that the equation for η should be a BSDE:

$$\begin{cases} d\eta(s) = \alpha(s)ds + \zeta(s)dW(s), \quad s \in [t, T], \\ \eta(T) = g, \end{cases}$$

where $\alpha : [t, T] \times \Omega \to \mathbb{R}^{2n}$ is to be determined. Applying Itô's formula to both sides of $Y = \Pi X + \eta$, we obtain

$$-\big(A^\top Y + C^\top Z + QI_nX + S^\top I_m u + q\big)ds + ZdW$$
$$= \big[\dot{\Pi}X + \Pi(AX + Bu + b) + \alpha\big]ds + \big[\Pi(CX + Du + \sigma) + \zeta\big]dW$$
$$= \big[(\dot{\Pi} + \Pi A)X + \Pi Bu + \Pi b + \alpha\big]ds + \big[\Pi CX + \Pi Du + \Pi\sigma + \zeta\big]dW.$$

Comparing the drift and diffusion coefficients, we find

$$\begin{aligned} A^\top Y + C^\top Z + QI_nX + S^\top I_m u + q \\ + (\dot{\Pi} + \Pi A)X + \Pi Bu + \Pi b + \alpha = 0, \end{aligned} \quad (2.2.5)$$

and

$$Z = \Pi CX + \Pi Du + \Pi\sigma + \zeta. \quad (2.2.6)$$

Substituting $Y = \Pi X + \eta$ and (2.2.6) into (2.2.4) yields

$$J^\top(B^\top \Pi + D^\top \Pi C + S I_n) X + J^\top(R I_m + D^\top \Pi D) u$$
$$+ J^\top(B^\top \eta + D^\top \zeta + D^\top \Pi \sigma + \rho) = 0. \tag{2.2.7}$$

If the $\mathbb{R}^{m \times m}$-valued function $J^\top(R I_m + D^\top \Pi D)$ has bounded inverse, then (2.2.7) implies

$$u = -\left[J^\top(R I_m + D^\top \Pi D)\right]^{-1} J^\top(B^\top \Pi + D^\top \Pi C + S I_n) X$$
$$- \left[J^\top(R I_m + D^\top \Pi D)\right]^{-1} J^\top(B^\top \eta + D^\top \zeta + D^\top \Pi \sigma + \rho), \tag{2.2.8}$$

which, together with (2.2.5) and (2.2.6), in turn yields

$$\begin{aligned}
0 &= A^\top(\Pi X + \eta) + C^\top(\Pi C X + \Pi D u + \Pi \sigma + \zeta) \\
&\quad + Q I_n X + S^\top I_m u + q + (\dot{\Pi} + \Pi A) X + \Pi B u + \Pi b + \alpha \\
&= (\dot{\Pi} + \Pi A + A^\top \Pi + C^\top \Pi C + Q I_n) X + (\Pi B + C^\top \Pi D + S^\top I_m) u \\
&\quad + A^\top \eta + C^\top \zeta + C^\top \Pi \sigma + \Pi b + q + \alpha \\
&= (\dot{\Pi} + \Pi A + A^\top \Pi + C^\top \Pi C + Q I_n) X \\
&\quad - (\Pi B + C^\top \Pi D + S^\top I_m) \left[J^\top(R I_m + D^\top \Pi D)\right]^{-1} \\
&\quad \times J^\top(B^\top \Pi + D^\top \Pi C + S I_n) X \\
&\quad - (\Pi B + C^\top \Pi D + S^\top I_m) \left[J^\top(R I_m + D^\top \Pi D)\right]^{-1} \\
&\quad \times J^\top(B^\top \eta + D^\top \zeta + D^\top \Pi \sigma + \rho) \\
&\quad + A^\top \eta + C^\top \zeta + C^\top \Pi \sigma + \Pi b + q + \alpha.
\end{aligned}$$

Comparing the coefficients of X in the above, we see that Π should be the solution to the following Riccati equation on $[t, T]$:

$$\begin{cases}
\dot{\Pi} + \Pi A + A^\top \Pi + C^\top \Pi C + Q I_n - (\Pi B + C^\top \Pi D + S^\top I_m) \\
\quad \times \left[J^\top(R I_m + D^\top \Pi D)\right]^{-1} J^\top(B^\top \Pi + D^\top \Pi C + S I_n) = 0, \tag{2.2.9} \\
\Pi(T) = G I_n,
\end{cases}$$

and that the BSDE for (η, ζ) should be

$$\begin{cases}
d\eta(s) = -\{(A^\top + K B^\top)\eta + (C^\top + K D^\top)\zeta + (C^\top + K D^\top)\Pi \sigma \\
\quad + \Pi b + q + K \rho\} ds + \zeta dW(s), \\
\eta(T) = g,
\end{cases}$$

where

$$K = -(\Pi B + C^\top \Pi D + S^\top I_m)\left[J^\top(R I_m + D^\top \Pi D)\right]^{-1} J^\top.$$

If the Riccati equation (2.2.9) indeed admits a solution Π such that $J^\top(RI_m + D^\top\Pi D)$ has bounded inverse, then by reversing the above argument, we see the 4-tuple (X, Y, Z, u), defined through the forward SDE

$$
\begin{cases}
dX = \left\{\left[A - B[J^\top(RI_m + D^\top\Pi D)]^{-1}J^\top(B^\top\Pi + D^\top\Pi C + SI_n)\right]X \right. \\
\qquad - B[J^\top(RI_m + D^\top\Pi D)]^{-1}J^\top(B^\top\eta + D^\top\zeta + D^\top\Pi\sigma + \rho) + b\Big\}ds \\
\qquad + \left\{\left[C - D[J^\top(RI_m + D^\top\Pi D)]^{-1}J^\top(B^\top\Pi + D^\top\Pi C + SI_n)\right]X \right. \\
\qquad \left. - D[J^\top(RI_m + D^\top\Pi D)]^{-1}J^\top(B^\top\eta + D^\top\zeta + D^\top\Pi\sigma + \rho) + \sigma\right\}dW, \\
X(t) = x,
\end{cases}
$$

and the Eqs. (2.2.8), $Y = \Pi X + \eta$, and (2.2.6), satisfies (2.2.3)–(2.2.4). If, in addition, the convexity condition (b) of Theorem 2.2.1 holds for $i = 1, 2$, then by Theorem 2.2.1, the control pair u defined by (2.2.8) is an open-loop Nash equilibrium for the initial state x. Observe that u takes the form

$$u(s) = \Theta(s)X(s) + v(s), \quad s \in [t, T],$$

where $(\Theta, v) \in \mathbf{\Theta}[t, T] \times \mathcal{U}[t, T]$ is independent of x. Writing (Θ, v) component-wise as

$$(\Theta, v) = (\Theta_1, v_1; \Theta_2, v_2) \quad \text{with } \Theta_i \in \mathbf{\Theta}_i[t, T], \ v_i \in \mathcal{U}_i[t, T]; \quad i = 1, 2,$$

we see then that (Θ, v) is a pair of closed-loop strategies. This leads to the following definition.

Definition 2.2.3 We say that the open-loop Nash equilibria of Problem (SDG) with initial time t admit a *closed-loop representation*, if there exists a pair $(\Theta, v) \in \mathbf{\Theta}[t, T] \times \mathcal{U}[t, T]$ of closed-loop strategies such that for any initial state $x \in \mathbb{R}^n$, the process

$$u(s) \triangleq \Theta(s)X(s) + v(s), \quad s \in [t, T] \tag{2.2.10}$$

is an open-loop Nash equilibrium of Problem (SDG) for (t, x), where $X(\cdot) = X(\cdot; t, x, \Theta, v)$ is the solution to the following closed-loop system:

$$
\begin{cases}
dX(s) = \{[A(s) + B(s)\Theta(s)]X(s) + B(s)v(s) + b(s)\}ds \\
\qquad + \{[C(s) + D(s)\Theta(s)]X(s) + D(s)v(s) + \sigma(s)\}dW(s), \tag{2.2.11} \\
X(t) = x.
\end{cases}
$$

Comparing Definitions 2.1.3 and 2.2.3, it is natural to ask: Does the closed-loop representation of open-loop Nash equilibria coincide with the outcome of some closed-loop Nash equilibrium? The answer is no. A counterexample will be presented in Sect. 2.4.

Now we present a characterization of the closed-loop representation of open-loop Nash equilibria.

Theorem 2.2.4 *Let* (G1)–(G2) *hold. Let* $(\Theta, v) \in \Theta[t, T] \times \mathcal{U}[t, T]$. *The open-loop Nash equilibria of Problem* (SDG) *with initial time t admit the closed-loop representation* (2.2.10) *if and only if*

(a) the convexity condition (b) of Theorem 2.2.1 holds for $i = 1, 2$,
(b) the solution $\Pi \in C([t, T]; \mathbb{R}^{n \times 2n})$ to the ODE on $[t, T]$

$$\begin{cases} \dot{\Pi} + \Pi A + A^\top \Pi + C^\top \Pi C + Q I_n \\ \quad + \left(\Pi B + C^\top \Pi D + S^\top I_m \right) \Theta = 0, \\ \Pi(T) = G I_n, \end{cases} \qquad (2.2.12)$$

satisfies

$$\left[J^\top (R I_m + D^\top \Pi D) \right] \Theta + J^\top \left(B^\top \Pi + D^\top \Pi C + S I_n \right) = 0 \qquad (2.2.13)$$

almost everywhere on $[t, T]$, and
(c) the adapted solution (η, ζ) to the BSDE on $[t, T]$

$$\begin{cases} d\eta(s) = -\left[A^\top \eta + C^\top \zeta + \left(\Pi B + C^\top \Pi D + S^\top I_m \right) v \right. \\ \quad \left. + C^\top \Pi \sigma + \Pi b + q \right] ds + \zeta dW(s), \\ \eta(T) = g, \end{cases} \qquad (2.2.14)$$

satisfies

$$\left[J^\top (R I_m + D^\top \Pi D) \right] v + J^\top \left(B^\top \eta + D^\top \zeta + D^\top \Pi \sigma + \rho \right) = 0 \qquad (2.2.15)$$

almost surely and almost everywhere on $[t, T]$.

Proof For arbitrary but fixed $x \in \mathbb{R}^n$, let X be the solution to (2.2.11). Let u be defined by (2.2.10) and set

$$Y = \Pi X + \eta, \quad Z = \Pi(C + D\Theta)X + \Pi Dv + \Pi \sigma + \zeta.$$

Then $Y(T) = G I_n X(T) + g$, and

$$dY = \dot{\Pi} X ds + \Pi dX + d\eta$$

$$= \big[\dot{\Pi} X + \Pi(A + B\Theta)X + \Pi Bv + \Pi b - A^\top \eta - C^\top \zeta$$
$$\quad - (\Pi B + C^\top \Pi D + S^\top I_m)v - C^\top \Pi \sigma - \Pi b - q\big]ds$$
$$\quad + \big[\Pi(C + D\Theta)X + \Pi Dv + \Pi \sigma + \zeta\big]dW$$

$$= \big[- (A^\top \Pi + C^\top \Pi C + QI_n + C^\top \Pi D\Theta + S^\top I_m \Theta)X$$
$$\quad - A^\top \eta - C^\top \zeta - (C^\top \Pi D + S^\top I_m)v - C^\top \Pi \sigma - q\big]ds + ZdW$$

$$= \big\{ - A^\top(\Pi X + \eta) - QI_n X - C^\top[\Pi(C + D\Theta)X + \Pi Dv + \Pi \sigma + \zeta]$$
$$\quad - S^\top I_m(\Theta X + v) - q\big\}ds + ZdW$$

$$= \big(- A^\top Y - QI_n X - C^\top Z - S^\top I_m u - q\big)ds + ZdW.$$

This shows that (X, Y, Z, u) satisfies the FBSDE (2.2.3). According to Theorem 2.2.1, the control pair u defined by (2.2.10) is an open-loop Nash equilibrium for (t, x) if and only if (a) holds and

$$0 = J^\top\big(B^\top Y + D^\top Z + SI_n X + RI_m u + \rho\big)$$
$$= J^\top\big\{B^\top(\Pi X + \eta) + D^\top[\Pi(C + D\Theta)X + \Pi Dv + \Pi \sigma + \zeta]$$
$$\quad + SI_n X + RI_m(\Theta X + v) + \rho\big\}$$
$$= J^\top\big[B^\top \Pi + D^\top \Pi C + SI_n + (RI_m + D^\top \Pi D)\Theta\big]X$$
$$\quad + J^\top\big[B^\top \eta + D^\top \zeta + D^\top \Pi \sigma + \rho + (RI_m + D^\top \Pi D)v\big]. \qquad (2.2.16)$$

Since the initial state x is arbitrary and $J^\top[B^\top \eta + D^\top \zeta + D^\top \Pi \sigma + \rho + (RI_m + D^\top \Pi D)v]$ is independent of x, we conclude that (2.2.16) is equivalent to (2.2.13) and (2.2.15). $\qquad \square$

We conclude this section with a remark on the solution Π of Eq. (2.2.12). The Eqs. (2.2.12) and (2.2.13) can be written componentwise as

$$\begin{cases} \dot{\Pi}_i + \Pi_i A + A^\top \Pi_i + C^\top \Pi_i C + Q^i \\ \quad + [\Pi_i B + C^\top \Pi_i D + (S^i)^\top]\Theta = 0, \\ \Pi_i(T) = G^i; \quad i = 1, 2, \end{cases} \qquad (2.2.17)$$

and

$$\begin{pmatrix} R_1^1 + D_1^\top \Pi_1 D \\ R_2^2 + D_2^\top \Pi_2 D \end{pmatrix} \Theta + \begin{pmatrix} B_1^\top \Pi_1 + D_1^\top \Pi_1 C + S_1^1 \\ B_2^\top \Pi_2 + D_2^\top \Pi_2 C + S_2^2 \end{pmatrix} = 0, \qquad (2.2.18)$$

respectively. The relation (2.2.18) shows that the equations for Π_1 and Π_2 are coupled. Since the $\mathbb{R}^{m \times m}$-valued function

$$\begin{pmatrix} R_1^1 + D_1^\top \Pi_1 D \\ R_2^2 + D_2^\top \Pi_2 D \end{pmatrix} = \begin{pmatrix} R_{11}^1 + D_1^\top \Pi_1 D_1 & R_{12}^1 + D_1^\top \Pi_1 D_2 \\ R_{21}^2 + D_2^\top \Pi_2 D_1 & R_{22}^2 + D_2^\top \Pi_2 D_2 \end{pmatrix}$$

is not necessarily symmetric even if D_1 and D_2 are zero, we see that, in general, Π_1 and Π_2 are not symmetric.

2.3 Closed-Loop Nash Equilibria and Symmetric Riccati Equations

In this section we characterize the closed-loop Nash equilibria of Problem (SDG). The idea is similar to that of the previous section: We regard the components of a closed-loop Nash equilibrium as closed-loop optimal strategies of certain LQ control problems. To make this idea precise, let $(\Theta_1^*, v_1^*; \Theta_2^*, v_2^*)$ be a pair of closed-loop strategies of Problem (SDG) on $[t, T]$, and consider the state equation

$$\begin{cases} dX(s) = [(A + B_2\Theta_2^*)X + B_1u_1 + B_2v_2^* + b]ds \\ \qquad\qquad + [(C + D_2\Theta_2^*)X + D_1u_1 + D_2v_2^* + \sigma]dW(s), \\ X(t) = x, \end{cases} \tag{2.3.1}$$

with the cost functional $\widehat{J}(t, x; u_1) \triangleq J^1(t, x; u_1, \Theta_2^*X + v_2^*)$, that is,

$$\begin{aligned} \widehat{J}(t, x; u_1) &= \mathbb{E}\Bigg\{ \langle G^1 X(T), X(T)\rangle + 2\langle g^1, X(T)\rangle \\ &\quad + \int_t^T \Bigg[\left\langle \begin{pmatrix} Q^1 & (S_1^1)^\top & (S_2^1)^\top \\ S_1^1 & R_{11}^1 & R_{12}^1 \\ S_2^1 & R_{21}^1 & R_{22}^1 \end{pmatrix} \begin{pmatrix} X \\ u_1 \\ \Theta_2^*X + v_2^* \end{pmatrix}, \begin{pmatrix} X \\ u_1 \\ \Theta_2^*X + v_2^* \end{pmatrix} \right\rangle \\ &\quad + 2\left\langle \begin{pmatrix} q^1 \\ \rho_1^1 \\ \rho_2^1 \end{pmatrix}, \begin{pmatrix} X \\ u_1 \\ \Theta_2^*X + v_2^* \end{pmatrix} \right\rangle \Bigg] ds \Bigg\} \\ &= \mathbb{E}\Bigg\{ \langle G^1 X(T), X(T)\rangle + 2\langle g^1, X(T)\rangle \\ &\quad + \int_t^T \Bigg[\left\langle \begin{pmatrix} \widehat{Q} & \widehat{S}^\top \\ \widehat{S} & R_{11}^1 \end{pmatrix} \begin{pmatrix} X \\ u_1 \end{pmatrix}, \begin{pmatrix} X \\ u_1 \end{pmatrix} \right\rangle + 2\left\langle \begin{pmatrix} \widehat{q} \\ \widehat{\rho} \end{pmatrix}, \begin{pmatrix} X \\ u_1 \end{pmatrix} \right\rangle \Bigg] ds \Bigg\} + \widehat{c}, \tag{2.3.2} \end{aligned}$$

where

$$\widehat{c} = \mathbb{E} \int_t^T [\langle R_{22}^1 v_2^*, v_2^*\rangle + 2\langle \rho_2^1, v_2^*\rangle]ds$$

is a constant and

$$\begin{aligned} \widehat{Q} &= Q^1 + (S_2^1)^\top\Theta_2^* + (\Theta_2^*)^\top S_2^1 + (\Theta_2^*)^\top R_{22}^1\Theta_2^*, & \widehat{S} &= S_1^1 + R_{12}^1\Theta_2^*, \\ \widehat{q} &= q^1 + (\Theta_2^*)^\top\rho_2^1 + (S_2^1 + R_{22}^1\Theta_2^*)^\top v_2^*, & \widehat{\rho} &= \rho_1^1 + R_{12}^1 v_2^*. \end{aligned}$$

We denote the problem of minimizing (2.3.2) subject to the Eq. (2.3.1) by Problem (1). In a similar fashion, we may consider the problem of minimizing

$$\widehat{J}(t, x; u_2) \triangleq J^2(t, x; \Theta_1^* X + v_1^*, u_2)$$

subject to the state equation

$$\begin{cases} dX(s) = [(A + B_1\Theta_1^*)X + B_2 u_2 + B_1 v_1^* + b]ds \\ \qquad\quad + [(C + D_1\Theta_1^*)X + D_2 u_2 + D_1 v_1^* + \sigma]dW(s), \\ X(t) = x, \end{cases}$$

which we denote by Problem (2).

Theorem 2.3.1 *Let* (G1)–(G2) *hold. Let* $(\Theta_1^*, v_1^*; \Theta_2^*, v_2^*) \in \boldsymbol{\Theta}_1[t, T] \times \mathcal{U}_1[t, T]$ $\times \boldsymbol{\Theta}_2[t, T] \times \mathcal{U}_2[t, T]$ *and denote*

$$\Theta^* = \begin{pmatrix} \Theta_1^* \\ \Theta_2^* \end{pmatrix}, \quad v^* = \begin{pmatrix} v_1^* \\ v_2^* \end{pmatrix}.$$

Then $(\Theta_1^*, v_1^*; \Theta_2^*, v_2^*)$ *is a closed-loop Nash equilibrium of Problem* (SDG) *on* $[t, T]$ *if and only if for* $i = 1, 2$,

(a) *the solution* $P_i \in C([t, T]; \mathbb{S}^n)$ *to the symmetric Lyapunov type equation*

$$\begin{cases} \dot{P}_i + P_i A + A^\top P_i + C^\top P_i C + Q^i + (\Theta^*)^\top (R^i + D^\top P_i D)\Theta^* \\ \quad + [P_i B + C^\top P_i D + (S^i)^\top]\Theta^* \\ \quad + (\Theta^*)^\top [B^\top P_i + D^\top P_i C + S^i] = 0, \\ P_i(T) = G^i \end{cases} \tag{2.3.3}$$

satisfies the following two conditions: for a.e. $s \in [t, T]$,

$$R_{ii}^i + D_i^\top P_i D_i \geqslant 0, \tag{2.3.4}$$

$$B_i^\top P_i + D_i^\top P_i C + S_i^i + (R_i^i + D_i^\top P_i D)\Theta^* = 0; \tag{2.3.5}$$

(b) *the adapted solution* (η_i, ζ_i) *to the BSDE on* $[t, T]$

$$\begin{cases} d\eta_i(s) = -\Big\{(A + B\Theta^*)^\top \eta_i + (C + D\Theta^*)^\top \zeta_i \\ \quad + [(\Theta^*)^\top (R^i + D^\top P_i D) + P_i B + C^\top P_i D + (S^i)^\top]v^* \\ \quad + (C + D\Theta^*)^\top P_i \sigma + (\Theta^*)^\top \rho^i + P_i b + q^i\Big\}ds + \zeta_i dW, \\ \eta_i(T) = g^i \end{cases} \tag{2.3.6}$$

satisfies the following condition: for a.e. $s \in [t, T]$,

$$B_i^\top \eta_i + D_i^\top \zeta_i + D_i^\top P_i \sigma + \rho_i^i + (R_i^i + D_i^\top P_i D)v^* = 0, \quad a.s. \qquad (2.3.7)$$

Proof From Proposition 2.1.5 we see that $(\Theta_1^*, v_1^*; \Theta_2^*, v_2^*)$ is a closed-loop Nash equilibrium of Problem (SDG) if and only if (Θ_1^*, v_1^*) is a closed-loop optimal strategy of Problem (1) and (Θ_2^*, v_2^*) is a closed-loop optimal strategy of Problem (2). According to Theorem 1.1.13 of Chap. 1, (Θ_1^*, v_1^*) is a closed-loop optimal strategy of Problem (1) if and only if the solution $P_1 \in C([t, T]; \mathbb{S}^n)$ to the ODE

$$\begin{cases} \dot{P}_1 + P_1(A + B_2\Theta_2^*) + (A + B_2\Theta_2^*)^\top P_1 + (C + D_2\Theta_2^*)^\top P_1(C + D_2\Theta_2^*) \\ \quad + \widehat{Q} + (\Theta_1^*)^\top (R_{11}^1 + D_1^\top P_1 D_1)\Theta_1^* \\ \quad + [P_1 B_1 + (C + D_2\Theta_2^*)^\top P_1 D_1 + \widehat{S}^\top]\Theta_1^* \\ \quad + (\Theta_1^*)[B_1^\top P_1 + D_1^\top P_1(C + D_2\Theta_2^*) + \widehat{S}] = 0, \\ P_1(T) = G^1 \end{cases}$$

is such that for a.e. $s \in [t, T]$,

$$R_{11}^1 + D_1^\top P_1 D_1 \geqslant 0,$$
$$B_1^\top P_1 + D_1^\top P_1(C + D_2\Theta_2^*) + \widehat{S} + (R_{11}^1 + D_1^\top P_1 D_1)\Theta_1^* = 0,$$

and the adapted solution (η_1, ζ_1) to the BSDE

$$\begin{cases} d\eta_1(s) = -\Big\{[(A + B_2\Theta_2^*) + B_1\Theta_1^*]^\top \eta_1 + [(C + D_2\Theta_2^*) + D_1\Theta_1^*]^\top \zeta_1 \\ \quad + [(C + D_2\Theta_2^*) + D_1\Theta_1^*]^\top P_1(D_2 v_2^* + \sigma) \\ \quad + (\Theta_1^*)^\top \hat{\rho} + P_1(B_2 v_2^* + b) + \hat{q}\Big\}ds + \zeta_1 dW(s), \\ \eta_1(T) = g^1 \end{cases}$$

is such that for a.e. $s \in [t, T]$,

$$B_1^\top \eta_1 + D_1^\top \zeta_1 + D_1^\top P_1(D_2 v_2^* + \sigma) + \hat{\rho} + (R_{11}^1 + D_1^\top P_1 D_1)v_1^* = 0, \quad a.s.$$

A tedious but straightforward calculation shows that the above equations can respectively be simplified to the following:

$$\begin{cases} \dot{P}_1 + P_1 A + A^\top P_1 + C^\top P_1 C + Q^1 + (\Theta^*)^\top (R^1 + D^\top P_1 D)\Theta^* \\ \quad + [P_1 B + C^\top P_1 D + (S^1)^\top]\Theta^* + (\Theta^*)^\top [B^\top P_1 + D^\top P_1 C + S^1] = 0, \\ P_1(T) = G^1, \end{cases}$$

$$R_{11}^1 + D_1^\top P_1 D_1 \geqslant 0,$$
$$B_1^\top P_1 + D_1^\top P_1 C + S_1^1 + (R_1^1 + D_1^\top P_1 D)\Theta^* = 0,$$

$$\begin{cases} d\eta_1(s) = -\Big\{ (A + B\Theta^*)^\top \eta_1 + (C + D\Theta^*)^\top \zeta_1 \\ \qquad + \big[(\Theta^*)^\top (R^1 + D^\top P_1 D) + P_1 B + C^\top P_1 D + (S^1)^\top \big] v^* \\ \qquad + (C + D\Theta^*)^\top P_1 \sigma + (\Theta^*)^\top \rho^1 + P_1 b + q^1 \Big\} ds + \zeta_1 dW(s), \\ \eta_1(T) = g^1, \\ B_1^\top \eta_1 + D_1^\top \zeta_1 + D_1^\top P_1 \sigma + \rho_1^1 + (R_1^1 + D_1^\top P_1 D) v^* = 0, \quad \text{a.s.} \end{cases}$$

From the above, we see that (Θ_1^*, v_1^*) is a closed-loop optimal strategy of Problem (1) if and only if the conditions (a) and (b) hold for $i = 1$. In a similar manner we can show that (Θ_2^*, v_2^*) is a closed-loop optimal strategy of Problem (2) if and only if the conditions (a) and (b) hold for $i = 2$. □

Note that conditions (2.3.5) and (2.3.7) are, respectively, equivalent to:

$$\begin{pmatrix} B_1^\top P_1 + D_1^\top P_1 C + S_1^1 \\ B_2^\top P_2 + D_2^\top P_2 C + S_2^2 \end{pmatrix} + \begin{pmatrix} R_1^1 + D_1^\top P_1 D \\ R_2^2 + D_2^\top P_2 D \end{pmatrix} \Theta^* = 0,$$

$$\begin{pmatrix} B_1^\top \eta_1 + D_1^\top \zeta_1 + D_1^\top P_1 \sigma + \rho_1^1 \\ B_2^\top \eta_2 + D_2^\top \zeta_2 + D_2^\top P_2 \sigma + \rho_2^2 \end{pmatrix} + \begin{pmatrix} R_1^1 + D_1^\top P_1 D \\ R_2^2 + D_2^\top P_2 D \end{pmatrix} v^* = 0.$$

Therefore,

$$\Theta^* = - \begin{pmatrix} R_1^1 + D_1^\top P_1 D \\ R_2^2 + D_2^\top P_2 D \end{pmatrix}^{-1} \begin{pmatrix} B_1^\top P_1 + D_1^\top P_1 C + S_1^1 \\ B_2^\top P_2 + D_2^\top P_2 C + S_2^2 \end{pmatrix}, \tag{2.3.8}$$

$$v^* = - \begin{pmatrix} R_1^1 + D_1^\top P_1 D \\ R_2^2 + D_2^\top P_2 D \end{pmatrix}^{-1} \begin{pmatrix} B_1^\top \eta_1 + D_1^\top \zeta_1 + D_1^\top P_1 \sigma + \rho_1^1 \\ B_2^\top \eta_2 + D_2^\top \zeta_2 + D_2^\top P_2 \sigma + \rho_2^2 \end{pmatrix},$$

provided the involved inverse (which is an $\mathbb{R}^{m \times m}$-valued function) exists. Plugging (2.3.8) into (2.3.3), we see that the equations for P_1 and P_2 are coupled, symmetric, and of Riccati type.

We conclude this section with a compact form of the Lyapunov type Eqs. (2.3.3) $(i = 1, 2)$. The differential equations in (2.3.3) $(i = 1, 2)$ can be expressed in block matrix notation as

$$0 = \begin{pmatrix} \dot{P}_1 & 0 \\ 0 & \dot{P}_2 \end{pmatrix} + \begin{pmatrix} P_1 & 0 \\ 0 & P_2 \end{pmatrix}\begin{pmatrix} A & 0 \\ 0 & A \end{pmatrix} + \begin{pmatrix} A & 0 \\ 0 & A \end{pmatrix}^{\top}\begin{pmatrix} P_1 & 0 \\ 0 & P_2 \end{pmatrix}$$

$$+ \begin{pmatrix} C & 0 \\ 0 & C \end{pmatrix}^{\top}\begin{pmatrix} P_1 & 0 \\ 0 & P_2 \end{pmatrix}\begin{pmatrix} C & 0 \\ 0 & C \end{pmatrix} + \begin{pmatrix} Q_1 & 0 \\ 0 & Q_2 \end{pmatrix}$$

$$+ \begin{pmatrix} \Theta^* & 0 \\ 0 & \Theta^* \end{pmatrix}^{\top}\begin{pmatrix} R^1 + D^{\top}P_1 D & 0 \\ 0 & R^2 + D^{\top}P_2 D \end{pmatrix}\begin{pmatrix} \Theta^* & 0 \\ 0 & \Theta^* \end{pmatrix}$$

$$+ \begin{pmatrix} P_1 B + C^{\top}P_1 D + (S^1)^{\top} & 0 \\ 0 & P_2 B + C^{\top}P_2 D + (S^2)^{\top} \end{pmatrix}\begin{pmatrix} \Theta^* & 0 \\ 0 & \Theta^* \end{pmatrix}$$

$$+ \begin{pmatrix} \Theta^* & 0 \\ 0 & \Theta^* \end{pmatrix}^{\top}\begin{pmatrix} B^{\top}P_1 + D^{\top}P_1 C + S^1 & 0 \\ 0 & B^{\top}P_2 + D^{\top}P_2 C + S^2 \end{pmatrix}.$$

Denoting

$$P \equiv \begin{pmatrix} P_1 & 0 \\ 0 & P_2 \end{pmatrix}, \quad \boldsymbol{\Theta} \equiv \begin{pmatrix} \Theta^* & 0 \\ 0 & \Theta^* \end{pmatrix}.$$

and using the notation introduced in the previous section, we can write

$$\begin{pmatrix} R^1 + D^{\top}P_1 D & 0 \\ 0 & R^2 + D^{\top}P_2 D \end{pmatrix} = R + D^{\top}PD,$$

$$\begin{pmatrix} P_1 B + C^{\top}P_1 D + (S^1)^{\top} & 0 \\ 0 & P_2 B + C^{\top}P_2 D + (S^2)^{\top} \end{pmatrix} = PB + C^{\top}PD + S^{\top}.$$

Consequently, one sees that the Eqs. (2.3.3) ($i = 1, 2$) are equivalent to

$$\begin{cases} \dot{P} + PA + A^{\top}P + C^{\top}PC + Q + \boldsymbol{\Theta}^{\top}(R + D^{\top}PD)\boldsymbol{\Theta} \\ \quad + (PB + C^{\top}PD + S^{\top})\boldsymbol{\Theta} + \boldsymbol{\Theta}^{\top}(B^{\top}P + D^{\top}PC + S) = 0, \\ P(T) = G, \end{cases}$$

which is easily seen to be symmetric. Note that in the notation of the previous section, the conditions (2.3.5) ($i = 1, 2$) can be rewritten as

$$0 = \begin{pmatrix} B_1^\top P_1 + D_1^\top P_1 C + S_1^1 \\ B_2^\top P_2 + D_2^\top P_2 C + S_2^2 \end{pmatrix} + \begin{pmatrix} R_1^1 + D_1^\top P_1 D \\ R_2^2 + D_2^\top P_2 D \end{pmatrix} \Theta^*$$

$$= \begin{pmatrix} B_1^\top & 0 \\ 0 & B_2^\top \end{pmatrix} \begin{pmatrix} P_1 & 0 \\ 0 & P_2 \end{pmatrix} \begin{pmatrix} I_n \\ I_n \end{pmatrix} + \begin{pmatrix} D_1^\top & 0 \\ 0 & D_2^\top \end{pmatrix} \begin{pmatrix} P_1 & 0 \\ 0 & P_2 \end{pmatrix} \begin{pmatrix} C & 0 \\ 0 & C \end{pmatrix} \begin{pmatrix} I_n \\ I_n \end{pmatrix}$$

$$+ \begin{pmatrix} I_{m_1} & 0 & 0 & 0 \\ 0 & 0 & 0 & I_{m_2} \end{pmatrix} \begin{pmatrix} S^1 & 0 \\ 0 & S^2 \end{pmatrix} \begin{pmatrix} I_n \\ I_n \end{pmatrix} + \begin{pmatrix} I_{m_1} & 0 & 0 & 0 \\ 0 & 0 & 0 & I_{m_2} \end{pmatrix} \begin{pmatrix} R^1 & 0 \\ 0 & R^2 \end{pmatrix} \begin{pmatrix} I_m \\ I_m \end{pmatrix} \Theta^*$$

$$+ \begin{pmatrix} D_1^\top & 0 \\ 0 & D_2^\top \end{pmatrix} \begin{pmatrix} P_1 & 0 \\ 0 & P_2 \end{pmatrix} \begin{pmatrix} D & 0 \\ 0 & D \end{pmatrix} \begin{pmatrix} I_m \\ I_m \end{pmatrix} \Theta^*$$

$$= \boldsymbol{J}^\top (\boldsymbol{B}^\top \boldsymbol{P} + \boldsymbol{D}^\top \boldsymbol{P} C + \boldsymbol{S}) I_n + [\boldsymbol{J}^\top (\boldsymbol{R} + \boldsymbol{D}^\top \boldsymbol{P} D) I_m] \Theta^*.$$

2.4 Relations Between Open-Loop and Closed-Loop Nash Equilibria

Recall that for the LQ optimal control problem, which is a special case of Problem (SDG), the closed-loop solvability implies the open-loop solvability. It is natural to expect that an analogous result holds for the LQ two-person differential game. However, as the next example shows, the existence of a closed-loop Nash equilibrium does not necessarily imply the existence of an open-loop Nash equilibrium. In this section, we discuss the connections between open-loop and closed-loop Nash equilibria by presenting three examples. It turns out that the open-loop and closed-loop solvability of Problem (SDG) are quite different.

The following example shows that Problem (SDG) may have only closed-loop Nash equilibria and no open-loop Nash equilibria.

Example 2.4.1 Consider the one-dimensional state equation

$$\begin{cases} dX(s) = u_1(s)ds + u_2(s)dW(s), & s \in [t, 1], \\ X(t) = x, \end{cases}$$

and the cost functionals

$$J^1(t, x; u_1, u_2) = \mathbb{E}\left\{|X(1)|^2 + \int_t^1 |u_1(s)|^2 ds\right\},$$

$$J^2(t, x; u_1, u_2) = \mathbb{E}\left\{-|X(1)|^2 + \int_t^1 \left[-|X(s)|^2 + |u_2(s)|^2\right] ds\right\}.$$

Let $t \in [0, 1)$ be arbitrary. We claim that

$$(\Theta_1^*(s), v_1^*(s); \Theta_2^*(s), v_2^*(s)) = ((s-2)^{-1}, 0; 0, 0)$$

is a closed-loop Nash equilibrium of the above problem on $[t, 1]$. According to Theorem 2.3.1, to verify this claim it suffices to show that for $i = 1, 2$, the solution P_i to the Eq. (2.3.3) satisfies (2.3.4) and (2.3.5), since in this example the solution (η_i, ζ_i) to the BSDE (2.3.6) is identically zero. Simplifying the equations for P_1 and P_2 we obtain

$$\begin{cases} \dot{P}_1(s) + P_1(s)\Theta_2^*(s)^2 + 2P_1(s)\Theta_1^*(s) + \Theta_1^*(s)^2 = 0, \\ P_1(1) = 1, \end{cases}$$

$$\begin{cases} \dot{P}_2(s) + P_2(s)\Theta_2^*(s)^2 + 2P_2(s)\Theta_1^*(s) + \Theta_2^*(s)^2 - 1 = 0, \\ P_2(1) = -1. \end{cases}$$

Substituting in $\Theta_1^*(s) = (s-2)^{-1}$ and $\Theta_2^*(s) = 0$ and solving for P_1 and P_2 yield

$$P_1(s) = \frac{1}{2-s}, \qquad P_2(s) = \frac{-(2-s)^3 - 2}{3(2-s)^2}.$$

Now one can easily verify that (2.3.4) and (2.3.5) hold.

Next we prove, by contradiction, that the problem does *not* have open-loop Nash equilibria. Suppose that (u_1^*, u_2^*) is an open-loop Nash equilibrium for some initial pair (t, x) with $t \in [0, 1)$. Then u_2^* is an open-loop optimal control of the following Problem (SLQ): To minimize

$$\widehat{J}(t, x; u_2) = \mathbb{E}\left\{-|X(1)|^2 + \int_t^1 \left[-|X(s)|^2 + |u_2(s)|^2\right] ds\right\} \qquad (2.4.1)$$

subject to the state equation

$$\begin{cases} dX(s) = u_1^*(s)ds + u_2(s)dW(s), \quad s \in [t, 1], \\ X(t) = x. \end{cases}$$

Take u_2 to be an arbitrary constant λ. Then the corresponding solution to the above equation is given by

$$X(s) = x + \int_t^s u_1^*(r)dr + \lambda[W(s) - W(t)]. \tag{2.4.2}$$

Let $\varepsilon > 0$ be undetermined. Substituting (2.4.2) into (2.4.1) and using the inequality $(a + b)^2 \geqslant (1 - \frac{1}{\varepsilon})a^2 + (1 - \varepsilon)b^2$, we obtain

$$\begin{aligned}
\widehat{J}(t, x; \lambda) &\leqslant \left(\frac{1}{\varepsilon} - 1\right)\mathbb{E}\left(x + \int_t^1 u_1^*(s)ds\right)^2 + (\varepsilon - 1)\mathbb{E}\left(\lambda[W(1) - W(t)]\right)^2 \\
&\quad + \left(\frac{1}{\varepsilon} - 1\right)\mathbb{E}\int_t^1\left(x + \int_t^s u_1^*(r)dr\right)^2 ds \\
&\quad + (\varepsilon - 1)\mathbb{E}\int_t^1\left(\lambda[W(s) - W(t)]\right)^2 ds + \lambda^2(1 - t) \\
&= \left(\frac{1}{\varepsilon} - 1\right)\mathbb{E}\left[\left(x + \int_t^1 u_1^*(s)ds\right)^2 + \int_t^1\left(x + \int_t^s u_1^*(r)dr\right)^2 ds\right] \\
&\quad + \frac{1}{2}\lambda^2(1 - t)\left[2\varepsilon + (\varepsilon - 1)(1 - t)\right].
\end{aligned}$$

Choosing $\varepsilon > 0$ small enough so that $2\varepsilon + (\varepsilon - 1)(1 - t) < 0$ and then letting $\lambda \to \infty$, we see that

$$\inf_{u_2 \in L_{\mathbb{F}}^2(t,1;\mathbb{R})} \widehat{J}(t, x; u_2) = -\infty,$$

which contradicts the fact that the LQ problem has an open-loop optimal control u_2^*.

The following example shows that Problem (SDG) may have only open-loop Nash equilibria and no closed-loop Nash equilibria.

Example 2.4.2 Consider the one-dimensional state equation

$$\begin{cases} dX(s) = [u_1(s) + u_2(s)]ds + [u_1(s) - u_2(s)]dW(s), & s \in [t, 1], \\ X(t) = x, \end{cases}$$

and the cost functionals

$$J^1(t, x; u_1, u_2) = J^2(t, x; u_1, u_2) = \mathbb{E}|X(1)|^2 \equiv J(t, x; u_1, u_2).$$

Let $t \in [0, 1)$ be arbitrary, and let $\beta \geqslant \frac{1}{1-t}$. We claim that

$$\left(u_1^\beta(s), u_2^\beta(s)\right) = -\left(\frac{\beta x}{2}\mathbf{1}_{[t,t+\frac{1}{\beta}]}(s), \frac{\beta x}{2}\mathbf{1}_{[t,t+\frac{1}{\beta}]}(s)\right), \quad s \in [t, 1]$$

is an open-loop Nash equilibrium for the initial pair (t, x). Indeed, it is clear that $J(t, x; u_1, u_2^\beta) \geqslant 0$ for any $u_1 \in L_\mathbb{F}^2(t, 1; \mathbb{R})$. On the other hand, the state process X^β corresponding to (u_1^β, u_2^β) and (t, x) satisfies $X^\beta(1) = 0$. Hence,

$$J(t, x; u_1^\beta, u_2^\beta) = 0 \leqslant J(t, x; u_1, u_2^\beta), \quad \forall u_1 \in L_\mathbb{F}^2(t, 1; \mathbb{R}).$$

Likewise,

$$J(t, x; u_1^\beta, u_2^\beta) = 0 \leqslant J(t, x; u_1^\beta, u_2), \quad \forall u_2 \in L_\mathbb{F}^2(t, 1; \mathbb{R}).$$

This establishes the claim.

However, the problem does *not* admit a closed-loop Nash equilibrium. This can be proved by contradiction. Suppose that $(\Theta_1^*, v_1^*; \Theta_2^*, v_2^*)$ is a closed-loop Nash equilibrium on $[t, 1]$ with $t < 1$. Consider the corresponding ODEs in Theorem 2.3.1, which now become

$$\begin{cases} \dot{P}_i + P_i(\Theta_1^* - \Theta_2^*)^2 + 2P_i(\Theta_1^* + \Theta_2^*) = 0, \\ P_i(1) = 1. \end{cases} \quad i = 1, 2. \tag{2.4.3}$$

The corresponding constraints read

$$P_1, P_2 \geqslant 0, \quad P_1 + P_1(\Theta_1^* - \Theta_2^*) = 0, \quad P_2 - P_2(\Theta_1^* - \Theta_2^*) = 0. \tag{2.4.4}$$

Since P_1 and P_2 satisfy the same ODE (2.4.3), we have $P_1 = P_2$ by the uniqueness of solutions. Then the last two equations in (2.4.4) imply $P_1 \equiv 0$, which contradicts the terminal condition $P_1(1) = 1$.

The next example shows that the closed-loop representation of open-loop Nash equilibria is not necessarily the outcome of a closed-loop Nash equilibrium.

Example 2.4.3 Consider the one-dimensional state equation

$$\begin{cases} dX(s) = [u_1(s) + u_2(s)]ds + X(s)dW(s), \quad s \in [t, T], \\ X(t) = x, \end{cases}$$

and the cost functionals

$$J^1(t, x; u_1, u_2) = \mathbb{E}\left[X(T)^2 + \int_t^T u_1(s)^2 ds\right],$$

$$J^2(t, x; u_1, u_2) = \mathbb{E}\left[X(T)^2 + \int_t^T u_2(s)^2 ds\right].$$

In this example,

$$A = 0, \ C = 1, \qquad B_1 = B_2 = 1, \qquad D_1 = D_2 = 0, \qquad b = \sigma = 0,$$

$$\rho^1 = \rho^2 = \begin{pmatrix} 0 \\ 0 \end{pmatrix}, \qquad S^1 = S^2 = \begin{pmatrix} 0 \\ 0 \end{pmatrix}, \qquad R^1 = \begin{pmatrix} 1 & 0 \\ 0 & 0 \end{pmatrix}, \qquad R^2 = \begin{pmatrix} 0 & 0 \\ 0 & 1 \end{pmatrix},$$

$$q^1 = q^2 = 0, \qquad Q^1 = Q^2 = 0, \qquad G^1 = G^2 = 1, \qquad g^1 = g^2 = 0.$$

Clearly, the convexity condition (b) of Theorem 2.2.1 holds for $i = 1, 2$. The corresponding Eqs. (2.2.12)–(2.2.13) can be written componentwise as

$$\dot{\Pi}_i(s) + \Pi_i(s) + \Pi_i(s)[\Theta_1(s) + \Theta_2(s)] = 0, \quad \Pi_i(T) = 1;$$
$$\Theta_i(s) + \Pi_i(s) = 0; \quad i = 1, 2.$$

It is not hard to see that the solutions Π_1 and Π_2 are equal and both are given by

$$\Pi_1(s) = \Pi_2(s) = \frac{e^{T-s}}{2e^{T-s} - 1}.$$

Note that if we take v to be zero, then the adapted solution (η, ζ) to the BSDE (2.2.14) is identically $(0, 0)$ and hence (2.2.15) holds. Thus, by Theorem 2.2.4, the open-loop Nash equilibria of this problem (with initial time t) admit a closed-loop representation which is given by

$$u_1(s) = u_2(s) = -\frac{e^{T-s}}{2e^{T-s} - 1} X(s), \qquad s \in [t, T]. \qquad (2.4.5)$$

Next we show that the problem admits a unique closed-loop Nash equilibrium $(\Theta_1^*, v_1^*; \Theta_2^*, v_2^*)$ on $[t, T]$. In light of Theorem 2.3.1, to determine $(\Theta_1^*, v_1^*; \Theta_2^*, v_2^*)$, we need to solve the constrained Riccati equations

$$\begin{cases} \dot{P}_1(s) + P_1(s) + \Theta_1^*(s)^2 + 2P_1(s)[\Theta_1^*(s) + \Theta_2^*(s)] = 0, \\ P_1(T) = 1, \quad P_1(s) + \Theta_1^*(s) = 0, \end{cases} \qquad (2.4.6)$$

$$\begin{cases} \dot{P}_2(s) + P_2(s) + \Theta_2^*(s)^2 + 2P_2(s)[\Theta_1^*(s) + \Theta_2^*(s)] = 0, \\ P_2(T) = 1, \quad P_2(s) + \Theta_2^*(s) = 0, \end{cases} \qquad (2.4.7)$$

as well as the constrained BSDEs

$$
\begin{cases}
d\eta_1(s) = -\big\{[\Theta_1^*(s) + \Theta_2^*(s)]\eta_1(s) + \zeta_1(s) + \Theta_1^*(s)v_1^*(s) \\
\qquad\qquad + P_1(s)[v_1^*(s) + v_2^*(s)]\big\}ds + \zeta_1(s)dW(s), \\
\eta_1(T) = 0, \quad \eta_1(s) + v_1^*(s) = 0,
\end{cases}
\tag{2.4.8}
$$

$$
\begin{cases}
d\eta_2(s) = -\big\{[\Theta_1^*(s) + \Theta_2^*(s)]\eta_2(s) + \zeta_2(s) + \Theta_2^*(s)v_2^*(s) \\
\qquad\qquad + P_2(s)[v_1^*(s) + v_2^*(s)]\big\}ds + \zeta_2(s)dW(s), \\
\eta_2(T) = 0, \quad \eta_2(s) + v_2^*(s) = 0.
\end{cases}
\tag{2.4.9}
$$

Using the relations $P_i(s) + \Theta_i^*(s) = 0;\ i = 1, 2$, we can further rewrite (2.4.6) and (2.4.7) as follows:

$$
\begin{cases}
\dot{P}_1(s) = P_1(s)^2 + 2P_1(s)P_2(s) - P_1(s), \\
P_1(T) = 1, \quad \Theta_1^*(s) = -P_1(s),
\end{cases}
$$

$$
\begin{cases}
\dot{P}_2(s) = P_2(s)^2 + 2P_2(s)P_1(s) - P_2(s), \\
P_2(T) = 1, \quad \Theta_2^*(s) = -P_2(s).
\end{cases}
$$

Now it is easily seen that

$$
P_1(s) = P_2(s) = \frac{e^{T-s}}{3e^{T-s} - 2}, \quad \Theta_1^*(s) = \Theta_2^*(s) = -\frac{e^{T-s}}{3e^{T-s} - 2}.
$$

To solve (2.4.8) and (2.4.9), we first use the relations $\eta_i(s) + v_i^*(s) = 0;\ i = 1, 2$ to rewrite them as

$$
\begin{cases}
d\eta_1(s) = -\big\{K(s)\eta_1(s) + \zeta_1(s) - K(s)[v_1^*(s) + v_2^*(s)]\big\}ds + \zeta_1(s)dW(s), \\
\eta_1(T) = 0, \quad v_1^*(s) = -\eta_1(s),
\end{cases}
$$

$$
\begin{cases}
d\eta_2(s) = -\big\{K(s)\eta_2(s) + \zeta_2(s) - K(s)[v_1^*(s) + v_2^*(s)]\big\}ds + \zeta_2(s)dW(s), \\
\eta_2(T) = 0, \quad v_2^*(s) = -\eta_2(s),
\end{cases}
$$

where $K(s) = \Theta_1^*(s)$. Since (η_1, ζ_1) and (η_2, ζ_2) satisfy the same BSDE, they must be equal. Substituting $v_1^*(s) = v_2^*(s) = -\eta_1(s)$ into the BSDE for (η_1, ζ_1) yields

$$
\begin{cases}
d\eta_1(s) = -[3K(s)\eta_1(s) + \zeta_1(s)]ds + \zeta_1(s)dW(s), \\
\eta_1(T) = 0,
\end{cases}
$$

from which we find $(\eta_1, \zeta_1) = (0, 0)$, and hence $v_1^*(s) = v_2^*(s) = 0$. Therefore, the unique closed-loop Nash equilibrium $(\Theta_1^*, v_1^*; \Theta_2^*, v_2^*)$ is given by

$$
\Theta_1^*(s) = \Theta_2^*(s) = -\frac{e^{T-s}}{3e^{T-s} - 2}, \quad v_1^*(s) = v_2^*(s) = 0.
\tag{2.4.10}
$$

Comparing (2.4.5) with (2.4.10), we see that the closed-loop representation of open-loop Nash equilibria is different from the outcome of the closed-loop Nash equilibrium.

Remark 2.4.4 From Example 2.4.3 we see that the closed-loop representation of open-loop Nash equilibria is not the outcome of a closed-loop Nash equilibrium in general. The reason is that the Riccati equation (2.3.3) for P_i is symmetric, whereas the Eq. (2.2.17) for Π_i is not.

2.5 Zero-Sum Games

In this section we consider LQ two-person zero-sum differential games in which one player's gain is the other's loss. In this case, the sum of the payoffs/costs of the players is always zero, i.e.,

$$J^1(t, x; u_1, u_2) + J^2(t, x; u_1, u_2) = 0, \quad \forall u_i \in \mathcal{U}_i[t, T], \ i = 1, 2.$$

Thus, we may assume, without loss of generality, that the weighting matrices in J^1 and J^2 have opposite signs, i.e.,

$$
\begin{aligned}
&G^1 + G^2 = 0, \quad g^1 + g^2 = 0, \quad Q^1 + Q^2 = 0, \quad q^1 + q^2 = 0, \\
&S_j^1 + S_j^2 = 0, \quad R_{jk}^1 + R_{jk}^2 = 0, \quad \rho_j^1 + \rho_j^2 = 0; \quad j, k = 1, 2.
\end{aligned}
\tag{2.5.1}
$$

To simplify the notation we shall denote $J(t, x; u_1, u_2) = J^1(t, x; u_1, u_2)$ and

$$
\begin{aligned}
&B = (B_1, B_2), \quad g = g^1, \quad q = q^1, \\
&D = (D_1, D_2), \quad G = G^1, \quad Q = Q^1, \quad R_{jk} = R_{jk}^1, \\
&R = \begin{pmatrix} R_{11} & R_{12} \\ R_{21} & R_{22} \end{pmatrix}, \quad S = \begin{pmatrix} S_1^1 \\ S_2^1 \end{pmatrix}, \quad \rho = \begin{pmatrix} \rho_1^1 \\ \rho_2^1 \end{pmatrix}, \quad u = \begin{pmatrix} u_1 \\ u_2 \end{pmatrix}.
\end{aligned}
$$

Then the state equation (2.1.1) and the cost functional $J^1(t, x; u_1, u_2)$ can be rewritten, respectively, as

$$
\begin{cases}
dX(s) = [A(s)X(s) + B(s)u(s) + b(s)]ds \\
\qquad\qquad + [C(s)X(s) + D(s)u(s) + \sigma(s)]dW(s), \\
X(t) = x,
\end{cases}
$$

and

$$J(t, x; u) = \mathbb{E}\Big\{\langle GX(T), X(T)\rangle + 2\langle g, X(T)\rangle$$

$$+ \int_t^T \Big[\Big\langle \begin{pmatrix} Q(s) & S(s)^\top \\ S(s) & R(s) \end{pmatrix} \begin{pmatrix} X(s) \\ u(s) \end{pmatrix}, \begin{pmatrix} X(s) \\ u(s) \end{pmatrix}\Big\rangle$$

$$+ 2\Big\langle \begin{pmatrix} q(s) \\ \rho(s) \end{pmatrix}, \begin{pmatrix} X(s) \\ u(s) \end{pmatrix}\Big\rangle\Big] ds\Big\}.$$

In terms of the single cost functional $J(t, x; u_1, u_2)$, the notions of open-loop and closed-loop saddle points can be represented as follows.

Definition 2.5.1 Let (2.5.1) hold.

(i) A pair $(u_1^*, u_2^*) \in \mathcal{U}_1[t, T] \times \mathcal{U}_2[t, T]$ is called an *open-loop saddle point* of Problem (SDG) for the initial pair $(t, x) \in [0, T) \times \mathbb{R}^n$ if

$$J(t, x; u_1^*, u_2) \leqslant J(t, x; u_1^*, u_2^*) \leqslant J(t, x; u_1, u_2^*),$$
$$\forall (u_1, u_2) \in \mathcal{U}_1[t, T] \times \mathcal{U}_2[t, T].$$

(ii) A 4-tuple $(\Theta_1^*, v_1^*; \Theta_2^*, v_2^*) \in \boldsymbol{\Theta}_1[t, T] \times \mathcal{U}_1[t, T] \times \boldsymbol{\Theta}_2[t, T] \times \mathcal{U}_2[t, T]$ is called a *closed-loop saddle point* of Problem (SDG) on $[t, T]$ if the inequalities

$$J(t, x; \Theta_1^* X + v_1^*, u_2) \leqslant J(t, x; \Theta_1^* X^* + v_1^*, \Theta_2^* X^* + v_2^*)$$
$$\leqslant J(t, x; u_1, \Theta_2^* X + v_2^*)$$

hold for all $x \in \mathbb{R}^n$ and $(u_1, u_2) \in \mathcal{U}_1[t, T] \times \mathcal{U}_2[t, T]$.

2.5.1 Characterizations of Open-Loop and Closed-Loop Saddle Points

In the case of zero-sum games, Theorems 2.2.1 and 2.3.1 can be reformulated in the following simpler ways.

Theorem 2.5.2 *Let* (G1)–(G2) *and* (2.5.1) *hold. Let* $(u_1^*, u_2^*) \in \mathcal{U}_1[t, T] \times \mathcal{U}_2[t, T]$ *and denote*

$$u^* = \begin{pmatrix} u_1^* \\ u_2^* \end{pmatrix}.$$

Then (u_1^*, u_2^*) *is an open-loop saddle point of Problem* (SDG) *for the initial pair* (t, x) *if and only if*

(a) the adapted solution (X^*, Y^*, Z^*) *to the FBSDE*

$$\begin{cases} dX^*(s) = (AX^* + Bu^* + b)ds + (CX^* + Du^* + \sigma)dW(s), \\ dY^*(s) = -\big(A^\top Y^* + C^\top Z^* + QX^* + S^\top u^* + q\big)ds + Z^* dW(s), \\ X^*(t) = x, \quad Y^*(T) = GX^*(T) + g, \end{cases}$$

satisfies the stationarity condition

$$B^\top Y^* + D^\top Z^* + SX^* + Ru^* + \rho = 0, \quad \text{a.e. on } [t, T], \text{ a.s.};$$

(b) the following convexity-concavity condition holds: For $i = 1, 2$,

$$(-1)^{i-1}\mathbb{E}\left\{ \langle GX_i(T), X_i(T)\rangle + \int_t^T \left\langle \begin{pmatrix} Q & (S_i)^\top \\ S_i & R_{ii} \end{pmatrix}\begin{pmatrix} X_i \\ u_i \end{pmatrix}, \begin{pmatrix} X_i \\ u_i \end{pmatrix}\right\rangle ds \right\} \geq 0$$

for all $u_i \in \mathcal{U}_i[t, T]$, where X_i is the solution to the SDE

$$\begin{cases} dX_i(s) = (AX_i + B_i u_i)ds + (CX_i + D_i u_i)dW(s), \quad s \in [t, T], \\ X_i(t) = 0, \end{cases}$$

(or equivalently, the cost functional $J(t, x; u_1, u_2)$ is convex in u_1 and concave in u_2).

Proof The result follows obviously from the fact that in the case of (2.5.1), the solutions $(Y_i^*, Z_i^*); i = 1, 2$, to the BSDE in (2.2.1) are mutual additive inverses. □

Theorem 2.5.3 Let (G1)–(G2) and (2.5.1) hold. Let $(\Theta_1^*, v_1^*; \Theta_2^*, v_2^*) \in \Theta_1[t, T] \times \mathcal{U}_1[t, T] \times \Theta_2[t, T] \times \mathcal{U}_2[t, T]$ and denote

$$\Theta^* = \begin{pmatrix} \Theta_1^* \\ \Theta_2^* \end{pmatrix}, \quad v^* = \begin{pmatrix} v_1^* \\ v_2^* \end{pmatrix}.$$

Then $(\Theta_1^*, v_1^*; \Theta_2^*, v_2^*)$ is a closed-loop saddle point of Problem (SDG) on $[t, T]$ if and only if

(a) the solution $P \in C([t, T]; \mathbb{S}^n)$ to the symmetric Lyapunov type equation

$$\begin{cases} \dot{P} + PA + A^\top P + C^\top PC + Q + (\Theta^*)^\top(R + D^\top PD)\Theta^* \\ \quad + (PB + C^\top PD + S^\top)\Theta^* + (\Theta^*)^\top(B^\top P + D^\top PC + S) = 0, \\ P(T) = G, \end{cases}$$

satisfies the following conditions: for a.e. $s \in [t, T]$,

$$R_{11} + D_1^\top PD_1 \geqslant 0, \quad R_{22} + D_2^\top PD_2 \leqslant 0, \tag{2.5.2}$$

$$B^\top P + D^\top PC + S + (R + D^\top PD)\Theta^* = 0; \tag{2.5.3}$$

(b) the adapted solution (η, ζ) to the BSDE on $[t, T]$

$$\begin{cases} d\eta(s) = -\big[(A + B\Theta^*)^\top \eta + (C + D\Theta^*)^\top \zeta \\ \qquad\qquad + (C + D\Theta^*)^\top P\sigma + (\Theta^*)^\top \rho + Pb + q\big]ds + \zeta dW, \quad (2.5.4) \\ \eta(T) = g, \end{cases}$$

satisfies the following condition: for a.e. $s \in [t, T]$,

$$B^\top \eta + D^\top \zeta + D^\top P\sigma + \rho + (R + D^\top PD)v^* = 0, \quad a.s.$$

Proof In the case of (2.5.1), the solutions $P_i; i = 1, 2$, to the ODE (2.3.3) are mutual additive inverses:

$$P_1(s) = -P_2(s), \quad s \in [t, T]. \tag{2.5.5}$$

It is clear then that the conditions (a) in Theorems 2.3.1 and 2.5.3 are equivalent. From the relations (2.5.1) and (2.5.5) we see that the solutions $(\eta_i, \zeta_i); i = 1, 2$, to the BSDE (2.3.6) are also mutual additive inverses. Moreover, by substituting (2.5.3) into (2.3.6), the equation for (η_1, ζ_1) reduces to (2.5.4). The equivalence between the conditions (b) in Theorems 2.3.1 and 2.5.3 then follows easily. $\square$

The following statement is an equivalent version of Theorem 2.5.3, which provides an explicit representation for the closed-loop saddle points.

Theorem 2.5.4 *Let* (G1)–(G2) *and* (2.5.1) *hold. Then Problem* (SDG) *admits a closed-loop saddle point on* $[t, T]$ *if and only if*

(a) the Riccati equation

$$\begin{cases} \dot{P} + PA + A^\top P + C^\top PC + Q - (PB + C^\top PD + S^\top) \\ \quad \times (R + D^\top PD)^\dagger (B^\top P + D^\top PC + S) = 0, \quad s \in [t, T], \quad (2.5.6) \\ P(T) = G \end{cases}$$

admits a solution $P \in C([t, T]; \mathbb{S}^n)$ *such that*

$$R_{11} + D_1^\top PD_1 \geqslant 0, \quad R_{22} + D_2^\top PD_2 \leqslant 0, \quad \text{a.e. on } [t, T], \tag{2.5.7}$$

$$\mathscr{R}(B^\top P + D^\top PC + S) \subseteq \mathscr{R}(R + D^\top PD), \quad \text{a.e. on } [t, T], \tag{2.5.8}$$

$$(R + D^\top PD)^\dagger (B^\top P + D^\top PC + S) \in L^2(t, T; \mathbb{R}^{m \times n}); \tag{2.5.9}$$

(b) the adapted solution (η, ζ) *of the BSDE*

$$\begin{cases} d\eta(s) = -\big[(A + B\widehat{\Theta})^\top \eta + (C + D\widehat{\Theta})\zeta + (C + D\widehat{\Theta})P\sigma \\ \qquad\qquad + \widehat{\Theta}^\top \rho + Pb + q\big]ds + \zeta dW(s), \quad s \in [t, T], \quad (2.5.10) \\ \eta(T) = g, \end{cases}$$

where $\widehat{\Theta} \equiv -(R + D^\top PD)^\dagger (B^\top P + D^\top PC + S)$, *satisfies*

$$B^\top \eta + D^\top \zeta + D^\top P\sigma + \rho \in \mathscr{R}(R + D^\top PD), \quad \text{a.e. on } [t, T], \ a.s.$$
$$(R + D^\top PD)^\dagger (B^\top \eta + D^\top \zeta + D^\top P\sigma + \rho) \in L^2_\mathbb{F}(t, T; \mathbb{R}^m).$$

In this case, the closed-loop saddle points are given by

$$\Theta^* = - (R + D^\top PD)^\dagger (B^\top P + D^\top PC + S)$$
$$+ [I - (R + D^\top PD)^\dagger (R + D^\top PD)]\Gamma, \tag{2.5.11}$$
$$v^* = - (R + D^\top PD)^\dagger (B^\top \eta + D^\top \zeta + D^\top P\sigma + \rho)$$
$$+ [I - (R + D^\top PD)^\dagger (R + D^\top PD)]\gamma, \tag{2.5.12}$$

where $\Gamma \in L^2(t, T; \mathbb{R}^{m \times n})$ and $\gamma \in L^2_\mathbb{F}(t, T; \mathbb{R}^m)$ are arbitrary.

Proof The proof of the equivalence of Theorems 2.5.3 and 2.5.4 is straightforward, using Proposition 1.3.1 of Chap. 1. $\qquad\square$

2.5.2 Relations Between Open-Loop and Closed-Loop Saddle Points

We have seen in Sect. 2.4 that the existence of open-loop Nash equilibria does not necessarily imply the existence of closed-loop Nash equilibria, and vice-versa. Even if both open-loop and closed-loop Nash equilibria exist, the outcome of a closed-loop Nash equilibrium is not necessarily the closed-loop representation of open-loop Nash equilibria (see Example 2.4.3). The situation is somewhat different in the zero-sum case. We shall see in this section that any closed-loop representation of open-loop saddle points coincides with the outcome of a closed-loop saddle point, as long as both exist.

Let $(\Theta, v) \in \boldsymbol{\Theta}[t, T] \times \mathcal{U}[t, T]$, and assume that the open-loop saddle points of Problem (SDG) on $[t, T]$ admit the closed-loop representation (2.2.10). Recalling (2.5.1) and the notation introduced right after (2.5.1), we may rewrite the Eqs. (2.2.17) $(i = 1, 2)$ for Π_1 and Π_2 as

$$\begin{cases} \dot{\Pi}_1 + \Pi_1 A + A^\top \Pi_1 + C^\top \Pi_1 C + Q + (\Pi_1 B + C^\top \Pi_1 D + S^\top)\Theta = 0, \\ \Pi_1(T) = G, \end{cases}$$

and

$$\begin{cases} \dot{\Pi}_2 + \Pi_2 A + A^\top \Pi_2 + C^\top \Pi_2 C - Q + (\Pi_2 B + C^\top \Pi_2 D - S^\top)\Theta = 0, \\ \Pi_2(T) = -G, \end{cases}$$

respectively. It is easily seen that $\Pi_1 = -\Pi_2$. Hence, denoting $\Pi \triangleq \Pi_1$, we may rewrite (2.2.18) as

$$\begin{pmatrix} R_{11} + D_1^\top \Pi D_1 & R_{12} + D_1^\top \Pi D_2 \\ -R_{21} - D_2^\top \Pi D_1 & -R_{22} - D_2^\top \Pi D_2 \end{pmatrix} \Theta + \begin{pmatrix} B_1^\top \Pi + D_1^\top \Pi C + S_1 \\ -B_2^\top \Pi - D_2^\top \Pi C - S_2 \end{pmatrix} = 0,$$

or equivalently,

$$(R + D^\top \Pi D)\Theta + B^\top \Pi + D^\top \Pi C + S = 0.$$

According to Proposition 1.3.1 of Chap. 1, the latter in turn is equivalent to

$$\mathscr{R}(B^\top \Pi + D^\top \Pi C + S) \subseteq \mathscr{R}(R + D^\top \Pi D), \quad \text{a.e. on } [t, T], \tag{2.5.13}$$

$$(R + D^\top \Pi D)^\dagger (B^\top \Pi + D^\top \Pi C + S) \in L^2(t, T; \mathbb{R}^{m \times n}), \tag{2.5.14}$$

and in this case Θ is given by

$$\Theta = -(R + D^\top \Pi D)^\dagger (B^\top \Pi + D^\top \Pi C + S)$$
$$+ [I - (R + D^\top \Pi D)^\dagger (R + D^\top \Pi D)]\Gamma, \tag{2.5.15}$$

where $\Gamma \in L^2(t, T; \mathbb{R}^{m \times n})$ is arbitrary. Upon substition of (2.5.15) into the equation for $\Pi = \Pi_1$, the latter becomes

$$\begin{cases} \dot{\Pi} + \Pi A + A^\top \Pi + C^\top \Pi C + Q - (\Pi B + C^\top \Pi D + S^\top) \\ \quad \times (R + D^\top \Pi D)^\dagger (B^\top \Pi + D^\top \Pi C + S) = 0, \quad s \in [t, T], \tag{2.5.16} \\ \Pi(T) = G. \end{cases}$$

Note that the Eq. (2.5.16) is symmetric. Now we write the solution (η, ζ) to the BSDE (2.2.14) into a vector form as

$$\eta = \begin{pmatrix} \eta_1 \\ \eta_2 \end{pmatrix}, \quad \zeta = \begin{pmatrix} \zeta_1 \\ \zeta_2 \end{pmatrix},$$

where each component is valued in $\mathbb{R}^n$. In a similar manner we can show that

(i) $(\eta_1, \zeta_1) = -(\eta_2, \zeta_2)$, and both satisfy

$$\begin{cases} d\eta(s) = -\big[(A + B\tilde{\Theta})^\top \eta + (C + D\tilde{\Theta})^\top \zeta + (C + D\tilde{\Theta})^\top \Pi \sigma \\ \quad + \tilde{\Theta}^\top \rho + \Pi b + q\big]ds + \zeta dW(s), \quad s \in [t, T], \tag{2.5.17} \\ \eta(T) = g, \end{cases}$$

where $\tilde{\Theta} = -(R + D^\top \Pi D)^\dagger (B^\top \Pi + D^\top \Pi C + S)$,

(ii) the constraint (2.2.15) is equivalent to

$$B^\top \eta + D^\top \zeta + D^\top \Pi \sigma + \rho \in \mathscr{R}(R + D^\top \Pi D), \quad \text{a.e. on } [t, T], \text{ a.s.} \quad (2.5.18)$$

$$(R + D^\top \Pi D)^\dagger (B^\top \eta + D^\top \zeta + D^\top \Pi \sigma + \rho) \in L^2_{\mathbb{F}}(t, T; \mathbb{R}^m), \quad (2.5.19)$$

(iii) and in this case v is given by

$$v = -(R + D^\top \Pi D)^\dagger (B^\top \eta + D^\top \zeta + D^\top \Pi \sigma + \rho) \\ + [I - (R + D^\top \Pi D)^\dagger (R + D^\top \Pi D)]\gamma,$$

where $\gamma \in L^2_{\mathbb{F}}(t, T; \mathbb{R}^m)$ is arbitrary.

We summarize these observations in the following theorem.

Theorem 2.5.5 *Let* (G1)–(G2) *and* (2.5.1) *hold. The open-loop saddle points of Problem* (SDG) *with initial time t admit a closed-loop representation if and only if the following hold:*

(a) *the convexity-concavity condition* (b) *of Theorem 2.5.2 holds;*
(b) *the Riccati equation* (2.5.16) *admits a solution $\Pi \in C([t, T]; \mathbb{S}^n)$ such that* (2.5.13)–(2.5.14) *hold, and the adapted solution (η, ζ) to the BSDE* (2.5.17) *satisfies* (2.5.18)–(2.5.19).

In this case, $\Theta(s)X(s) + v(s)$; $s \in [t, T]$ is a closed-loop representation of open-loop saddle points if and only if

$$\Theta = -(R + D^\top \Pi D)^\dagger (B^\top \Pi + D^\top \Pi C + S) \\ + [I - (R + D^\top \Pi D)^\dagger (R + D^\top \Pi D)]\Gamma, \\ v = -(R + D^\top \Pi D)^\dagger (B^\top \eta + D^\top \zeta + D^\top \Pi \sigma + \rho) \\ + [I - (R + D^\top \Pi D)^\dagger (R + D^\top \Pi D)]\gamma,$$

for some $\Gamma \in L^2(t, T; \mathbb{R}^{m \times n})$ and $\gamma \in L^2_{\mathbb{F}}(t, T; \mathbb{R}^m)$.

Proof The result can be proved by combining Theorem 2.2.4 and the previous argument. We leave the details to the interested reader. □

Comparing Theorems 2.5.4 and 2.5.5, one may wonder whether the closed-loop representation of open-loop saddle points coincide with the outcome of closed-loop saddle points when both exist. The answer to this question is affirmative, as shown by the following result.

Theorem 2.5.6 *Let* (G1)–(G2) *and* (2.5.1) *hold. Suppose that both open-loop and closed-loop saddle points exist on $[t, T]$. If the open-loop saddle points admit a closed-loop representation, then this representation must be the outcome of a closed-loop saddle point.*

Proof The proof is immediate from Theorems 2.5.4 and 2.5.5, once we show that the solution P to the Riccati equation (2.5.6) with constraints (2.5.7)–(2.5.9) coincides with the solution Π to (2.5.16) with constraints (2.5.13)–(2.5.14). To this end, we fix an arbitrary $t' \in [t, T]$.

First, we note that if the convexity-concavity condition (b) of Theorem 2.5.2 holds for the initial time t, it also holds for t'. Indeed, for any $u_1 \in \mathcal{U}_1[t', T]$, let X_1 be the solution to

$$\begin{cases} dX_1(s) = (AX_1 + B_1u_1)ds + (CX_1 + D_1u_1)dW(s), & s \in [t', T], \\ X_1(t') = 0, \end{cases}$$

and define the *zero-extension* of u_1 as follows:

$$[01_{[t,t')} \oplus u_1](s) = \begin{cases} 0, & s \in [t, t'), \\ u_1(s), & s \in [t', T]. \end{cases}$$

Then $\tilde{u}_1 \equiv [01_{[t,t')} \oplus u_1] \in \mathcal{U}_1[t, T]$, and due to the initial state being 0, the solution $\tilde{X}_1$ of

$$\begin{cases} d\tilde{X}_1(s) = (A\tilde{X}_1 + B_1\tilde{u}_1)ds + (C\tilde{X}_1 + D_1\tilde{u}_1)dW(s), & s \in [t, T], \\ \tilde{X}_1(t) = 0 \end{cases}$$

is such that

$$\tilde{X}_1(s) = \begin{cases} 0, & s \in [t, t'), \\ X_1(s), & s \in [t', T]. \end{cases}$$

It follows that

$$\mathbb{E}\left\{ \langle GX_1(T), X_1(T) \rangle + \int_{t'}^T \left\langle \begin{pmatrix} Q & (S_1)^\top \\ S_1 & R_{11} \end{pmatrix} \begin{pmatrix} X_1 \\ u_1 \end{pmatrix}, \begin{pmatrix} X_1 \\ u_1 \end{pmatrix} \right\rangle ds \right\}$$
$$= \mathbb{E}\left\{ \langle G\tilde{X}_1(T), \tilde{X}_1(T) \rangle + \int_t^T \left\langle \begin{pmatrix} Q & (S_1)^\top \\ S_1 & R_{11} \end{pmatrix} \begin{pmatrix} \tilde{X}_1 \\ \tilde{u}_1 \end{pmatrix}, \begin{pmatrix} \tilde{X}_1 \\ \tilde{u}_1 \end{pmatrix} \right\rangle ds \right\} \geqslant 0.$$

This proves the case $i = 1$. The case $i = 2$ can be treated similarly.

Now let (η_p, ζ_p) denote the adapted solution to (2.5.10). It is clear that over the interval $[t', T]$, with Π replaced by P and (η, ζ) replaced by (η_p, ζ_p), the condition (b) of Theorem 2.5.5 still holds. It follows by Theorem 2.5.5 that if (Θ^*, v^*) is a closed-loop saddle point on $[t, T]$, then the outcome

$$u^*(s) = \Theta^*(s)X^*(s) + v^*(s), \quad s \in [t', T]$$

is a closed-loop representation of open-loop saddle points, where X^* is the solution of

$$\begin{cases} dX^*(s) = [(A + B\Theta^*)X^* + Bv^* + b]ds \\ \qquad\qquad + [(C + D\Theta^*)X^* + Dv^* + \sigma]dW(s), \quad s \in [t', T], \\ X^*(t') = x, \end{cases}$$

with arbitrary initial state $x \in \mathbb{R}^n$. Using the representation (2.5.11)–(2.5.12) of (Θ^*, v^*), one can easily verify that P satisfies

$$\dot{P} + P(A + B\Theta^*) + (A + B\Theta^*)^\top P + (C + D\Theta^*)^\top P(C + D\Theta^*)$$
$$+ (\Theta^*)^\top R\Theta^* + S^\top \Theta^* + (\Theta^*)^\top S + Q = 0,$$

and that (η_p, ζ_p) satisfies

$$d\eta_p = -\big[(A + B\Theta^*)^\top \eta_p + (C + D\Theta^*)^\top \zeta_p + (C + D\Theta^*)^\top P\sigma$$
$$+ (\Theta^*)^\top \rho + Pb + q\big]ds + \zeta_p dW.$$

Then applying Itô's formula to $s \mapsto \langle P(s)X^*(s), X^*(s) \rangle$ yields

$$\mathbb{E}\langle GX^*(T), X^*(T) \rangle - \mathbb{E}\langle P(t')x, x \rangle$$
$$= \mathbb{E}\int_{t'}^T \Big\{ -\langle [(\Theta^*)^\top R\Theta^* + S^\top \Theta^* + (\Theta^*)^\top S + Q]X^*, X^* \rangle$$
$$+ 2\langle P(Bv^* + b) + (C + D\Theta^*)^\top P(Dv^* + \sigma), X^* \rangle$$
$$+ \langle D^\top PDv^*, v^* \rangle + 2\langle D^\top P\sigma, v^* \rangle + \langle P\sigma, \sigma \rangle \Big\}ds,$$

and applying Itô's formula to $s \mapsto \langle \eta_p(s), X^*(s) \rangle$ yields

$$\mathbb{E}\langle g, X^*(T) \rangle - \mathbb{E}\langle \eta_p(t'), x \rangle$$
$$= \mathbb{E}\int_{t'}^T \Big\{ -\langle (C + D\Theta^*)^\top P\sigma + (\Theta^*)^\top \rho + Pb + q, X^* \rangle$$
$$+ \langle B^\top \eta_p + D^\top \zeta_p, v^* \rangle + \langle \eta_p, b \rangle + \langle \zeta_p, \sigma \rangle \Big\}ds.$$

Substituting for $\mathbb{E}\langle GX^*(T), X^*(T) \rangle$ and $\mathbb{E}\langle g, X^*(T) \rangle$ in the cost functional

$$J(t', x; u^*) = J(t', x; \Theta^* X^* + v^*)$$
$$= \mathbb{E}\Big\{ \langle GX^*(T), X^*(T) \rangle + 2\langle g, X^*(T) \rangle$$
$$+ \int_{t'}^T \Big[\langle [Q + S^\top \Theta^* + (\Theta^*)^\top S + (\Theta^*)^\top R\Theta^*]X^*, X^* \rangle$$
$$+ 2\langle (R\Theta^* + S)^\top v^* + (\Theta^*)^\top \rho + q, X^* \rangle$$
$$+ \langle Rv^*, v^* \rangle + 2\langle \rho, v^* \rangle \Big]ds \Big\}$$

and noting that

$$(R + D^\top P D)\Theta^* + B^\top P + D^\top P C + S = 0,$$

we obtain

$$
\begin{aligned}
J(t', x; u^*) = \mathbb{E}\Big\{ &\langle P(t')x, x \rangle + 2\langle \eta_p(t'), x \rangle \\
&+ \int_{t'}^{T} \Big[\langle P\sigma, \sigma \rangle + 2\langle \eta_p, b \rangle + 2\langle \zeta_p, \sigma \rangle + \langle (R + D^\top P D)v^*, v^* \rangle \\
&+ 2\langle B^\top \eta_p + D^\top \zeta_p + D^\top P\sigma + \rho, v^* \rangle \Big] ds \Big\}.
\end{aligned}
\tag{2.5.20}
$$

Next let (η_n, ζ_n) denote the adapted solution to (2.5.17), and let

$$
\begin{aligned}
\Theta &= -(R + D^\top \Pi D)^\dagger (B^\top \Pi + D^\top \Pi C + S), \\
v &= -(R + D^\top \Pi D)^\dagger (B^\top \eta_n + D^\top \zeta_n + D^\top \Pi\sigma + \rho).
\end{aligned}
$$

According to Theorem 2.5.5, for the initial time t',

$$u(s) = \Theta(s)X(s) + v(s), \quad s \in [t', T]$$

is a closed-loop representation of open-loop saddle points. We may proceed as previously to obtain

$$
\begin{aligned}
J(t', x; u) = \mathbb{E}\Big\{ &\langle \Pi(t')x, x \rangle + 2\langle \eta_n(t'), x \rangle \\
&+ \int_{t'}^{T} \Big[\langle \Pi\sigma, \sigma \rangle + 2\langle \eta_n, b \rangle + 2\langle \zeta_n, \sigma \rangle + \langle (R + D^\top \Pi D)v, v \rangle \\
&+ 2\langle B^\top \eta_n + D^\top \zeta_n + D^\top \Pi\sigma + \rho, v \rangle \Big] ds \Big\}.
\end{aligned}
\tag{2.5.21}
$$

Since both $u^* = \begin{pmatrix} u_1^* \\ u_2^* \end{pmatrix}$ and $u = \begin{pmatrix} u_1 \\ u_2 \end{pmatrix}$ are open-loop saddle points for (t', x), we have

$$
\begin{aligned}
J(t', x; u_1^*, u_2^*) &\leqslant J(t', x; u_1, u_2^*) \leqslant J(t', x; u_1, u_2) \\
&\leqslant J(t', x; u_1^*, u_2) \leqslant J(t', x; u_1^*, u_2^*),
\end{aligned}
$$

which implies $J(t', x; u^*) = J(t', x; u)$. Since x is arbitrary, we conclude from (2.5.20) and (2.5.21) that $P(t') = \Pi(t')$. $\qquad\square$

Remark 2.5.7 Theorem 2.5.6 is based on the assumption that both the closed-loop representation of open-loop saddle points and the closed-loop saddle points exist on $[t, T]$. This assumption is necessary because, in general, neither of these two kinds of existence implies the other (see Examples 2.5.8 and 2.5.9 below). It is different from Problem (SLQ), in which closed-loop solvability always implies open-loop solvability. Recall from Theorem 1.1.12 of Chap. 1 that for Problem (SLQ), when a closed-loop optimal strategy exists, the solution P to the associated Riccati equation satisfies $R + D^\top P D \geqslant 0$. This positivity condition actually implies the convexity condition (i) of Theorem 1.1.9 in Chap. 1. However, in the case of Problem (SDG), one cannot deduce the convexity-concavity condition (b) of Theorem 2.5.2 from the counterpart (2.5.7) of $R + D^\top P D \geqslant 0$, and vice-versa.

We now present an example to show that the existence of a closed-loop saddle point does not necessarily imply the existence of an open-loop saddle point.

Example 2.5.8 Consider the one-dimensional state equation

$$\begin{cases} dX(s) = [u_1(s) - u_2(s)]ds + [u_1(s) - u_2(s)]dW(s), & s \in [t, 1], \\ X(t) = x, \end{cases}$$

and the cost functional

$$J(t, x; u_1, u_2) = \mathbb{E}\left\{ |X(1)|^2 + \int_t^1 \left[|u_1(s)|^2 - |u_2(s)|^2 \right]ds \right\}.$$

The associated Riccati equation reads

$$\begin{cases} \dot{P}(s) = P(s)(1, -1) \begin{pmatrix} 1 + P(s) & -P(s) \\ -P(s) & -1 + P(s) \end{pmatrix}^\dagger \begin{pmatrix} 1 \\ -1 \end{pmatrix} P(s) = 0, \\ P(1) = 1. \end{cases}$$

One can easily check that $P(s) = 1$ $(0 \leqslant s \leqslant 1)$ is the unique solution. Since

$$R(s) + D(s)^\top P(s) D(s) = \begin{pmatrix} 2 & -1 \\ -1 & 0 \end{pmatrix}$$

is invertible, the range inclusion condition (2.5.8) automatically holds. Also note that

$$[R(s) + D(s)^\top P(s) D(s)]^\dagger B(s)^\top P(s) = \begin{pmatrix} 1 \\ 1 \end{pmatrix},$$

$$R_{11}(s) + D_1(s)^\top P(s) D_1(s) = 2, \quad R_{22}(s) + D_2(s)^\top P(s) D_2(s) = 0.$$

Hence the conditions (2.5.7) and (2.5.9) hold. The associated BSDE (2.5.10) reads

$$d\eta(s) = \zeta(s)dW(s), \quad s \in [t, 1]; \quad \eta(1) = 0,$$

whose solution is clearly $(0, 0)$, and hence the condition (b) of Theorem 2.5.4 is satisfied. Therefore, by Theorem 2.5.4, the game admits a unique closed-loop saddle point (Θ^*, v^*) over $[t, 1]$, which is given by

$$\Theta^*(s) = -\begin{pmatrix} 1 \\ 1 \end{pmatrix}, \quad v^*(s) = \begin{pmatrix} 0 \\ 0 \end{pmatrix}; \quad s \in [t, 1].$$

We next prove that the game does not have open-loop saddle points by showing that the convexity-concavity condition (b) of Theorem 2.5.2 fails. To this end, we take u_2 to be an arbitrary constant $\lambda \neq 0$. The solution to the SDE

$$\begin{cases} dX_2(s) = -\lambda ds - \lambda dW(s), & s \in [t, 1], \\ X_2(t) = 0 \end{cases}$$

is given by

$$X(s) = -\lambda(s - t) - \lambda[W(s) - W(t)],$$

and hence

$$\mathbb{E}\left\{ |X_2(1)|^2 - \int_t^1 |u_2(s)|^2 ds \right\} = \lambda^2(1 - t)^2 > 0.$$

This means that the convexity-concavity condition (b) of Theorem 2.5.2 does not hold for $i = 2$.

The next example shows that the existence of an open-loop saddle point does not necessarily imply the existence of a closed-loop saddle point.

Example 2.5.9 Consider the two-dimensional controlled state equation

$$d\begin{pmatrix} X_1(s) \\ X_2(s) \end{pmatrix} = \begin{pmatrix} u_1(s) \\ u_2(s) \end{pmatrix} ds, \quad s \in [t, T]; \quad \begin{pmatrix} X_1(t) \\ X_2(t) \end{pmatrix} = \begin{pmatrix} x_1 \\ x_2 \end{pmatrix},$$

and the cost functional

$$J(t, x; u_1, u_2) = \mathbb{E}\left[|X_1(T)|^2 - |X_2(T)|^2 \right].$$

Let $(t, x) \in [0, T) \times \mathbb{R}^2$ be an arbitrary initial pair with $x = (x_1, x_2)^\top$. Choose constants $\lambda_i \geqslant \frac{1}{T-t}$; $i = 1, 2$, and define

$$u_i^*(s) = -\lambda_i x_i \mathbf{1}_{[t, t + \frac{1}{\lambda_i}]}(s), \quad s \in [t, T]; \quad i = 1, 2.$$

It is straightforward to verify that for any $(u_1, u_2) \in L^2_{\mathbb{F}}(t, T; \mathbb{R}) \times L^2_{\mathbb{F}}(t, T; \mathbb{R})$

$$J(t, x; u_1^*, u_2) \leqslant J(t, x; u_1^*, u_2^*) = 0 \leqslant J(t, x; u_1, u_2^*).$$

Thus, (u_1^*, u_2^*) is an open-loop saddle point for (t, x). However, there is no closed-loop saddle point for this problem. Indeed, the associated Riccati equation reads

$$\dot{P}(s) = 0, \quad s \in [t, T]; \qquad P(T) = G = \begin{pmatrix} 1 & 0 \\ 0 & -1 \end{pmatrix}.$$

Clearly, the solution to the above equation is

$$P(s) = G, \quad t \leqslant s \leqslant T.$$

Since in this example,

$$B^\top P + D^\top PC + S = P = G, \quad R + D^\top PD = 0,$$

we see that the inclusion relation (2.5.8) does not hold. Consequently, the closed-loop saddle point does not exist by Theorem 2.5.4.

2.6 Differential Games in Infinite Horizons

We now look at an infinite-horizon problem. As before, we let $(\Omega, \mathcal{F}, \mathbb{F}, \mathbb{P})$ be a complete filtered probability space on which a one-dimensional standard Brownian motion $W = \{W(t); t \geqslant 0\}$ is defined with $\mathbb{F} = \{\mathcal{F}_t\}_{t \geqslant 0}$ being the usual augmentation of the natural filtration generated by W. Recall the following notation:

$$L_\mathbb{F}^2(\mathbb{R}^n) = \left\{ \varphi : [0, \infty) \times \Omega \to \mathbb{H} \mid \varphi \in \mathbb{F}, \mathbb{E} \int_0^\infty |\varphi(t)|^2 dt < \infty \right\},$$

$$\mathcal{X}_{loc}[0, \infty) = \left\{ \varphi : [0, \infty) \times \Omega \to \mathbb{R}^n \mid \varphi \in \mathbb{F} \text{ is continuous,} \right.$$

$$\left. \text{and } \mathbb{E}\left[\sup_{0 \leqslant t \leqslant T} |\varphi(t)|^2 \right] < \infty \text{ for every } T > 0 \right\},$$

$$\mathcal{X}[0, \infty) = \left\{ \varphi \in \mathcal{X}_{loc}[0, \infty) \mid \mathbb{E} \int_0^\infty |\varphi(t)|^2 dt < \infty \right\}.$$

For brevity, we write

$$\mathcal{U}_i = L_\mathbb{F}^2(\mathbb{R}^{m_i}), \quad i = 1, 2; \qquad \mathcal{U} = \mathcal{U}_1 \times \mathcal{U}_2.$$

Consider the following controlled linear SDE on $[0, \infty)$:

$$\begin{cases} dX(t) = [AX(t) + B_1 u_1(t) + B_2 u_2(t) + b(t)]dt \\ \qquad\qquad + [CX(t) + D_1 u_1(t) + D_2 u_2(t) + \sigma(t)]dW(t), \qquad (2.6.1) \\ X(0) = x, \end{cases}$$

where we assume the following (which is comparable with (G1)):

(G1)′ The coefficients and the nonhomogeneous terms of (2.6.1) satisfy

$$A, C \in \mathbb{R}^{n \times n}; \quad B_i, D_i \in \mathbb{R}^{n \times m_i}, \ i = 1, 2; \quad b, \sigma \in L^2_{\mathbb{F}}(\mathbb{R}^n).$$

Clearly, under (G1)′, for any $x \in \mathbb{R}^n$ and $(u_1, u_2) \in \mathcal{U}_1 \times \mathcal{U}_2$, (2.6.1) admits a unique solution $X(\cdot) \equiv X(\cdot\,; x, u_1, u_2) \in \mathcal{X}_{loc}[0, \infty)$, which is the state process. In this section, we are only concerned with zero-sum games. So Player 1 and Player 2 share the same performance functional:

$$J(x; u_1, u_2) \triangleq \mathbb{E} \int_0^\infty \left[\left\langle \begin{pmatrix} Q & S_1^\top & S_2^\top \\ S_1 & R_{11} & R_{12} \\ S_2 & R_{21} & R_{22} \end{pmatrix} \begin{pmatrix} X(t) \\ u_1(t) \\ u_2(t) \end{pmatrix}, \begin{pmatrix} X(t) \\ u_1(t) \\ u_2(t) \end{pmatrix} \right\rangle \right.$$
$$\left. + 2 \left\langle \begin{pmatrix} q(t) \\ \rho_1(t) \\ \rho_2(t) \end{pmatrix}, \begin{pmatrix} X(t) \\ u_1(t) \\ u_2(t) \end{pmatrix} \right\rangle \right] dt, \tag{2.6.2}$$

for which we impose the following assumption:

(G2)′ The weighting coefficients in (2.6.2) satisfy

$$Q \in \mathbb{S}^n, \quad R_{21}^\top = R_{12} \in \mathbb{R}^{m_1 \times m_2}, \quad q \in L^2_{\mathbb{F}}(\mathbb{R}^n),$$
$$S_i \in \mathbb{R}^{m_i \times n}, \quad R_{ii} \in \mathbb{S}^{m_i}, \quad \rho_i \in L^2_{\mathbb{F}}(\mathbb{R}^{m_i}); \quad i = 1, 2.$$

Note that in general, for $(x, u_1, u_2) \in \mathbb{R}^n \times \mathcal{U}_1 \times \mathcal{U}_2$, the solution $X(\cdot) \equiv X(\cdot\,; x, u_1, u_2)$ of (2.6.1) might just be in $\mathcal{X}_{loc}[0, \infty)$ and the performance functional $J(x; u_1, u_2)$ might not be defined. Therefore, we introduce the following set:

$$\mathcal{U}_{ad}(x) \triangleq \{ (u_1, u_2) \in \mathcal{U} \mid X(\cdot\,; x, u_1, u_2) \in \mathcal{X}[0, \infty) \}.$$

An element $(u_1, u_2) \in \mathcal{U}_{ad}(x)$ is called an *admissible control pair* for the initial state x, and the corresponding state process $X(\cdot) \equiv X(\cdot\,; x, u_1, u_2)$ is called an *admissible state process* with initial state x.

In the current zero-sum game, we assume that Player 1 is the minimizer and Player 2 is the maximizer. That is, Player 1 wishes to minimize (2.6.2) by selecting a control u_1, and Player 2 wishes to maximize (2.6.2) by selecting a control u_2. Thus, (2.6.2) represents the cost of Player 1 and the payoff of Player 2. The problem is to find an admissible control pair (u_1^*, u_2^*) that both players can accept. We refer to such a problem as an *infinite-horizon linear-quadratic stochastic two-person zero-sum differential game*, and denote it by Problem (SDG)$_\infty$ for short.

Similar to the finite-horizon problem, for notational convenience we let $m = m_1 + m_2$ and denote

$$B = (B_1, B_2), \quad D = (D_1, D_2),$$

$$R = \begin{pmatrix} R_{11} & R_{12} \\ R_{21} & R_{22} \end{pmatrix}, \quad S = \begin{pmatrix} S_1 \\ S_2 \end{pmatrix}, \quad \rho = \begin{pmatrix} \rho_1 \\ \rho_2 \end{pmatrix}, \quad u = \begin{pmatrix} u_1 \\ u_2 \end{pmatrix}.$$

With the above notation, the state equation can be rewritten as

$$\begin{cases} dX(t) = [A(t)X(t) + B(t)u(t) + b(t)]dt \\ \qquad\qquad + [C(t)X(t) + D(t)u(t) + \sigma(t)]dW(t), \quad t \geqslant 0, \\ X(0) = x, \end{cases}$$

and the performance functional can be rewritten as

$$J(x; u) = \mathbb{E} \int_0^\infty \left[\left\langle \begin{pmatrix} Q & S^{\mathsf{T}} \\ S & R \end{pmatrix} \begin{pmatrix} X(t) \\ u(t) \end{pmatrix}, \begin{pmatrix} X(t) \\ u(t) \end{pmatrix} \right\rangle + 2 \left\langle \begin{pmatrix} q(t) \\ \rho(t) \end{pmatrix}, \begin{pmatrix} X(t) \\ u(t) \end{pmatrix} \right\rangle \right] dt.$$

Also, when $b, \sigma, q, \rho = 0$, we denote the corresponding Problem (SDG)$_\infty$ by Problem (SDG)$_\infty^0$ and the corresponding performance functional by $J^0(x; u)$. Similar to Problem (SLQ)$_\infty$, we will assume that the set of stabilizers of the system

$$dX(t) = [AX(t) + Bu(t)]dt + [CX(t) + Du(t)]dW(t), \quad t \geqslant 0 \qquad (2.6.3)$$

is nonempty, that is,

$$\mathscr{S}[A, C; B, D] \triangleq \{\Theta \in \mathbb{R}^{m \times n} \mid \Theta \text{ stabilizes the system } (2.6.3)\} \neq \varnothing.$$

Moreover, for $\Theta_i \in \mathbb{R}^{m_i \times n}, i = 1, 2$, we let

$$\mathscr{S}_1(\Theta_2) = \left\{ \Theta_1 \in \mathbb{R}^{m_1 \times n} : \begin{pmatrix} \Theta_1 \\ \Theta_2 \end{pmatrix} \in \mathscr{S}[A, C; B, D] \right\},$$

$$\mathscr{S}_2(\Theta_1) = \left\{ \Theta_2 \in \mathbb{R}^{m_2 \times n} : \begin{pmatrix} \Theta_1 \\ \Theta_2 \end{pmatrix} \in \mathscr{S}[A, C; B, D] \right\}.$$

Note that in general, say, $\mathscr{S}_1(\Theta_2)$ is not necessarily non-empty for some $\Theta_2 \in \mathbb{R}^{m_2 \times n}$. However, if $\Theta \triangleq (\Theta_1^{\mathsf{T}}, \Theta_2^{\mathsf{T}})^{\mathsf{T}} \in \mathscr{S}[A, C; B, D]$, then both $\mathscr{S}_1(\Theta_2)$ and $\mathscr{S}_2(\Theta_1)$ are non-empty.

Definition 2.6.1 For a given initial state $x \in \mathbb{R}^n$, a pair $(\bar{u}_1, \bar{u}_2) \in \mathcal{U}_{ad}(x)$ is called an *open-loop saddle point* of Problem (SDG)$_\infty$ if

$$J(x; \bar{u}_1, u_2) \leqslant J(x; \bar{u}_1, \bar{u}_2) \leqslant J(x; u_1, \bar{u}_2)$$

for any $(u_1, u_2) \in \mathcal{U}$ such that $J(x; \bar{u}_1, u_2)$ and $J(x; u_1, \bar{u}_2)$ are defined.

Definition 2.6.2 A 4-tuple $(\Theta_1^*, u_1^*; \Theta_2^*, u_2^*) \in \mathbb{R}^{m_1 \times n} \times \mathcal{U}_1 \times \mathbb{R}^{m_2 \times n} \times \mathcal{U}_2$ is called a *closed-loop saddle point* of Problem (SDG)$_\infty$ if

(i) $\Theta^* \triangleq ((\Theta_1^*)^\top, (\Theta_2^*)^\top)^\top \in \mathscr{S}[A, C; B, D]$, and

(ii) for any $x \in \mathbb{R}^n$, $(\Theta_1, \Theta_2) \in \mathscr{S}_1(\Theta_2^*) \times \mathscr{S}_2(\Theta_1^*)$ and $(u_1, u_2) \in \mathcal{U}_1 \times \mathcal{U}_2$,

$$J(x; \Theta_1^* X + u_1^*, \Theta_2 X + u_2) \leqslant J(x; \Theta_1^* X^* + u_1^*, \Theta_2^* X^* + u_2^*)$$
$$\leqslant J(x; \Theta_1 X + u_1, \Theta_2^* X + u_2^*).$$

Remark 2.6.3

(a) Although both players are non-cooperative, when choosing Θ_i ($i = 1, 2$), they prefer to at least work together so that $\Theta = ((\Theta_1)^\top, (\Theta_2)^\top)^\top$ is a stabilizer of (2.6.3) (and the system will not be crashed). Thus, in Definition 2.6.2, we only require Θ^* being a stabilizer of (2.6.3) rather than Θ_i^* ($i = 1, 2$) being a stabilizer of the system

$$dX(t) = [AX(t) + B_i u_i(t)]dt + [CX(t) + D_i u_i(t)]dW(t), \quad t \geqslant 0.$$

(b) By a similar method used in proving Proposition 2.1.5 of [48, Chap. 2], one can show that the condition (ii) in Definition 2.6.2 is equivalent to the following:

(ii)' for any $x \in \mathbb{R}^n$ and $(u_1, u_2) \in \mathcal{U}_1 \times \mathcal{U}_2$,

$$J(x; \Theta_1^* X + u_1^*, \Theta_2^* X + u_2) \leqslant J(x; \Theta_1^* X^* + u_1^*, \Theta_2^* X^* + u_2^*)$$
$$\leqslant J(x; \Theta_1^* X + u_1, \Theta_2^* X + u_2^*).$$

Let

$$\Theta^* = \begin{pmatrix} \Theta_1^* \\ \Theta_2^* \end{pmatrix} \in \mathscr{S}[A, C; B, D], \quad u^* = \begin{pmatrix} u_1^* \\ u_2^* \end{pmatrix} \in \mathcal{U}.$$

Consider the state equation

$$\begin{cases} dX(t) = [(A + B\Theta^*)X(t) + Bu(t) + b(t)]dt \\ \qquad\qquad + [(C + D\Theta^*)X(t) + Du(t) + \sigma(t)]dW(t), \quad t \geqslant 0, \\ X(0) = x \end{cases}$$

and the performance functional

$$\tilde{J}(x; u_1, u_2) \triangleq J(x; \Theta_1^* X + u_1, \Theta_2^* X + u_2)$$
$$= \mathbb{E} \int_0^\infty \left[\left\langle \begin{pmatrix} \tilde{Q} & \tilde{S}^\top \\ \tilde{S} & R \end{pmatrix} \begin{pmatrix} X \\ u \end{pmatrix}, \begin{pmatrix} X \\ u \end{pmatrix} \right\rangle + 2 \left\langle \begin{pmatrix} \tilde{q} \\ \rho \end{pmatrix}, \begin{pmatrix} X \\ u \end{pmatrix} \right\rangle \right] dt,$$

where

$$\tilde{Q} = Q + (\Theta^*)^\top S + S^\top \Theta^* + (\Theta^*)^\top R\Theta^*,$$
$$\tilde{S} = S + R\Theta^*, \quad \tilde{q} = q + (\Theta^*)^\top \rho.$$

From (ii)$'$ of Remark 2.6.3, we see that $(\Theta_1^*, u_1^*; \Theta_2^*, u_2^*)$ is a closed-loop saddle point of Problem (SDG)$_\infty$ if and only if (u_1^*, u_2^*) is an open-loop saddle point for the problem with the above state equation and performance functional. Applying the idea used in the proof of Theorem 2.5.2, we see that $(\Theta_1^*, u_1^*; \Theta_2^*, u_2^*)$ is a closed-loop saddle point of Problem (SDG)$_\infty$ if and only if for any $x \in \mathbb{R}^n$, the adapted solution $(X^*, Y^*, Z^*) \in \mathcal{X}[0, \infty) \times \mathcal{X}[0, \infty) \times L_{\mathbb{F}}^2(\mathbb{R}^n)$ to the FBSDE

$$
\begin{cases}
dX^*(t) = \big[(A + B\Theta^*)X^* + Bu^* + b\big]dt \\
\qquad\quad + \big[(C + D\Theta^*)X^* + Du^* + \sigma\big]dW(t), \quad t \geqslant 0, \\
dY^*(t) = -\big[(A + B\Theta^*)^\top Y^* + (C + D\Theta^*)^\top Z^* \\
\qquad\quad + \widetilde{Q}X^* + \widetilde{S}^\top u^* + \tilde{q}\,\big]dt + Z^* dW(t), \quad t \geqslant 0, \\
X^*(0) = x
\end{cases}
\tag{2.6.4}
$$

satisfies the following stationarity condition:

$$
Ru^* + B^\top Y^* + D^\top Z^* + \widetilde{S}X^* + \rho = 0, \quad \text{a.e. a.s.}
$$

and the following condition holds for $i = 1, 2$:

$$
(-1)^{i-1}\mathbb{E} \int_0^\infty \left\langle \begin{pmatrix} \widetilde{Q} & \widetilde{S}_i^\top \\ \widetilde{S}_i & R_{ii} \end{pmatrix} \begin{pmatrix} X_i \\ u_i \end{pmatrix}, \begin{pmatrix} X_i \\ u_i \end{pmatrix} \right\rangle dt \geqslant 0, \quad \forall u_i \in \mathcal{U}_i,
$$

where $\widetilde{S}_i = S_i + R_i\Theta^*$ and X_i is the solution of

$$
\begin{cases}
dX_i(t) = [(A + B\Theta^*)X_i(t) + B_i u_i(t)]dt \\
\qquad\quad + [(C + D\Theta^*)X_i(t) + D_i u_i(t)]dW(t), \quad t \geqslant 0, \\
X_i(0) = 0.
\end{cases}
$$

By subtracting solutions of (2.6.4) with initial states x and 0, we obtain the following result.

Proposition 2.6.4 *If* $(\Theta_1^*, u_1^*; \Theta_2^*, u_2^*)$ *is a closed-loop saddle point of Problem* (SDG)$_\infty$, *then* $(\Theta_1^*, 0; \Theta_2^*, 0)$ *is a closed-loop saddle point of Problem* (SDG)$_\infty^0$.

Now we introduce the following algebraic Riccati equation (ARE):

$$
\begin{cases}
\mathcal{Q}(P) - \mathcal{S}(P)^\top \mathcal{R}(P)^\dagger \mathcal{S}(P) = 0, \\
\mathscr{R}(\mathcal{S}(P)) \subseteq \mathscr{R}(\mathcal{R}(P)), \\
\mathcal{R}_1(P) \geqslant 0, \quad \mathcal{R}_2(P) \leqslant 0,
\end{cases}
\tag{2.6.5}
$$

where

$$\mathcal{Q}(P) = PA + A^\mathsf{T}P + C^\mathsf{T}PC + Q,$$
$$\mathcal{S}(P) = B^\mathsf{T}P + D^\mathsf{T}PC^\mathsf{T} + S,$$
$$\mathcal{R}(P) = R + D^\mathsf{T}PD,$$
$$\mathcal{R}_i(P) = R_{ii} + D_i^\mathsf{T}PD_i; \quad i = 1, 2.$$

Definition 2.6.5 A matrix $P \in \mathbb{S}^n$ is called a *stabilizing solution* of (2.6.5) if P satisfies (2.6.5) and there exists a matrix $\Pi \in \mathbb{R}^{m \times n}$ such that

$$-\mathcal{R}(P)^\dagger \mathcal{S}(P) + [I - \mathcal{R}(P)^\dagger \mathcal{R}(P)]\Pi \in \mathscr{S}[A, C; B, D].$$

The following result provides a necessary condition for the existence of a closed-loop saddle point.

Proposition 2.6.6 *Suppose that Problem* $(SDG)_\infty^0$ *admits a closed-loop saddle point. Then the ARE (2.6.5) admits a stabilizing solution P.*

Proof We assume without loss of generality that $(\Theta_1^*, 0; \Theta_2^*, 0)$ is a closed-loop saddle point of Problem $(SDG)_\infty^0$. It is easily seen that the function $V^0(x)$ defined by

$$V^0(x) \triangleq J^0(x; \Theta_1^* X^*, \Theta_2^* X^*)$$

is a quadratic form, that is, there exists a matrix $P \in \mathbb{S}^n$ such that

$$V^0(x) = \langle Px, x \rangle, \quad \forall x \in \mathbb{R}^n.$$

Consider the state equation

$$\begin{cases} dX_1(t) = [(A + B_2\Theta_2^*)X_1(t) + B_1 u_1(t)]dt \\ \qquad\qquad + [(C + D_2\Theta_2^*)X_1(t) + D_1 u_1(t)]dW(t), \quad t \geq 0, \\ X_1(0) = x \end{cases}$$

and the cost functional

$$J_1(x; u_1) \triangleq J^0(x; u_1, \Theta_2^* X_1)$$
$$= \mathbb{E} \int_0^\infty \left\langle \begin{pmatrix} Q & S_1^\mathsf{T} & S_2^\mathsf{T} \\ S_1 & R_{11} & R_{12} \\ S_2 & R_{21} & R_{22} \end{pmatrix} \begin{pmatrix} X_1 \\ u_1 \\ \Theta_2^* X_1 \end{pmatrix}, \begin{pmatrix} X_1 \\ u_1 \\ \Theta_2^* X_1 \end{pmatrix} \right\rangle dt$$
$$= \mathbb{E} \int_0^\infty \left\{ \langle [Q + (\Theta_2^*)^\mathsf{T} R_{22}\Theta_2^* + (\Theta_2^*)^\mathsf{T} S_2 + S_2^\mathsf{T}\Theta_2^*]X_1, X_1 \rangle \right.$$
$$\left. + \langle R_{11} u_1, u_1 \rangle + 2\langle (S_1 + R_{12}\Theta_2^*)X_1, u_1 \rangle \right\} dt.$$

Then $(\Theta_1^*, 0)$ is a closed-loop optimal control of Problem $(SLQ)_\infty^0$ with the above state equation and cost functional, and the value function of this Problem $(SLQ)_\infty^0$ is given by $\langle Px, x \rangle$ with P satisfying

$$P\tilde{A}_1 + \tilde{A}_1^\top P + \tilde{C}_1^\top P \tilde{C}_1 + \tilde{Q}_1$$
$$- (PB_1 + \tilde{C}_1^\top PD_1 + \tilde{S}_1^\top)\mathcal{R}_1(P)^\dagger(B_1^\top P + D_1^\top P\tilde{C}_1 + \tilde{S}_1) = 0, \qquad (2.6.6)$$
$$\mathcal{R}_1(P) \geqslant 0, \quad \mathcal{R}_1(P)\Theta_1^* + (B_1^\top P + D_1^\top P\tilde{C}_1 + \tilde{S}_1) = 0, \qquad (2.6.7)$$

where

$$\tilde{A}_1 = A + B_2\Theta_2^*, \quad \tilde{C}_1 = C + D_2\Theta_2^*, \quad \mathcal{R}_1(P) = R_{11} + D_1^\top PD_1,$$
$$\tilde{Q}_1 = Q + (\Theta_2^*)^\top R_{22}\Theta_2^* + (\Theta_2^*)^\top S_2 + S_2^\top \Theta_2^*, \quad \tilde{S}_1 = S_1 + R_{12}\Theta_2^*.$$

Similarly, by considering the state equation

$$\begin{cases} dX_2(t) = [(A + B_1\Theta_1^*)X_2(t) + B_2u_2(t)]dt \\ \qquad\qquad + [(C + D_1\Theta_1^*)X_2(t) + D_2u_2(t)]dW(t), \quad t \geqslant 0, \\ X_2(0) = x \end{cases}$$

and the cost functional

$$J_2(x; u_2) \triangleq -J^0(x; \Theta_1^* X_2, u_2),$$

we can obtain

$$\mathcal{R}_2(P) \leqslant 0, \quad \mathcal{R}_2(P)\Theta_2^* + (B_2^\top P + D_2^\top P\tilde{C}_2 + \tilde{S}_2) = 0, \qquad (2.6.8)$$

where

$$\mathcal{R}_2(P) = R_{22} + D_2^\top PD_2, \quad \tilde{C}_2 = C + D_1\Theta_1^*, \quad \tilde{S}_2 = S_2 + R_{21}\Theta_1^*.$$

Let $\Theta^* = ((\Theta_1^*)^\top, (\Theta_2^*)^\top)^\top$. Combining (2.6.7) and (2.6.8), one has

$$\mathcal{R}(P)\Theta^* + \mathcal{S}(P) = (R + D^\top PD)\Theta^* + (B^\top P + D^\top PC + S) = 0, \qquad (2.6.9)$$

from which we conclude that

$$\mathscr{R}(\mathcal{S}(P)) \subseteq \mathscr{R}(\mathcal{R}(P))$$

and the existence of a matrix $\Pi \in \mathbb{R}^{m \times n}$ such that

$$\Theta^* = -\mathcal{R}(P)^\dagger \mathcal{S}(P) + [I - \mathcal{R}(P)^\dagger \mathcal{R}(P)]\Pi \in \mathscr{S}[A, C; B, D]. \qquad (2.6.10)$$

Using (2.6.6)–(2.6.9), we have

$$
\begin{aligned}
0 &= P\tilde{A}_1 + \tilde{A}_1^\top P + \tilde{C}_1^\top P\tilde{C}_1 + \tilde{Q}_1 \\
&\quad - (PB_1 + \tilde{C}_1^\top PD_1 + \tilde{S}_1^\top)\mathcal{R}_1(P)^\dagger(B_1^\top P + D_1^\top P\tilde{C}_1 + \tilde{S}_1) \\
&= P\tilde{A}_1 + \tilde{A}_1^\top P + \tilde{C}_1^\top P\tilde{C}_1 + \tilde{Q}_1 - (\Theta_1^*)^\top \mathcal{R}_1(P)\Theta_1^* \\
&= PA + A^\top P + C^\top PC + Q + (\Theta_2^*)^\top \mathcal{R}_2(P)\Theta_2^* - (\Theta_1^*)^\top \mathcal{R}_1(P)\Theta_1^* \\
&\quad + (PB_2 + C^\top PD_2 + S_2^\top)\Theta_2^* + (\Theta_2^*)^\top (B_2^\top P + D_2^\top PC + S_2) \\
&= \mathcal{Q}(P) - (\Theta_1^*)^\top \mathcal{R}_1(P)\Theta_1^* - (\Theta_2^*)^\top \mathcal{R}_2(P)\Theta_2^* \\
&\quad + \left[(\Theta_2^*)^\top \mathcal{R}_2(P) + (PB_2 + C^\top PD_2 + S_2^\top)\right]\Theta_2^* \\
&\quad + (\Theta_2^*)^\top \left[(B_2^\top P + D_2^\top PC + S_2) + \mathcal{R}_2(P)\Theta_2^*\right] \\
&= \mathcal{Q}(P) - (\Theta_1^*)^\top \mathcal{R}_1(P)\Theta_1^* - (\Theta_2^*)^\top \mathcal{R}_2(P)\Theta_2^* \\
&\quad - (\Theta_1^*)^\top (D_1^\top PD_2 + R_{12})\Theta_2^* - (\Theta_2^*)^\top (D_2^\top PD_1 + R_{21})\Theta_1^* \\
&= \mathcal{Q}(P) - (\Theta^*)^\top \mathcal{R}(P)\Theta^* \\
&= \mathcal{Q}(P) - \mathcal{S}(P)^\top \mathcal{R}(P)^\dagger \mathcal{S}(P).
\end{aligned}
$$

Therefore, P is a stabilizing solution of the ARE (2.6.5). $\square$

The following result, which is the main result of this section, gives a characterization for closed-loop saddle points of Problem (SDG)$_\infty$.

Theorem 2.6.7 *Problem* (SDG)$_\infty$ *admits a closed-loop saddle point*

$$
(\Theta^*, u^*) = \left(\begin{pmatrix} \Theta_1^* \\ \Theta_2^* \end{pmatrix}, \begin{pmatrix} u_1^* \\ u_2^* \end{pmatrix} \right) \in \mathbb{R}^{m\times n} \times \mathcal{U} \tag{2.6.11}
$$

if and only if

(i) *the ARE* (2.6.5) *admits a stabilizing solution* P, *and*
(ii) *the BSDE*

$$
\begin{aligned}
d\eta(t) = -\Big\{ &[A - B\mathcal{R}(P)^\dagger \mathcal{S}(P)]^\top \eta + [C - D\mathcal{R}(P)^\dagger \mathcal{S}(P)]^\top \zeta \\
&+ [C - D\mathcal{R}(P)^\dagger \mathcal{S}(P)]^\top P\sigma - [\mathcal{R}(P)^\dagger \mathcal{S}(P)]^\top \rho \\
&+ Pb + q \Big\} dt + \zeta dW(t), \quad t \geqslant 0,
\end{aligned} \tag{2.6.12}
$$

admits an L^2-stable adapted solution (η, ζ) such that

$$
B^\top \eta(t) + D^\top \zeta(t) + D^\top P\sigma(t) + \rho(t) \in \mathcal{R}(\mathcal{R}(P)), \quad \text{a.e. } t \in [0, \infty), \text{ a.s.}
$$

In this case, the closed-loop saddle point (Θ^*, u^*) admits the following representation:

$$
\begin{cases}
\Theta^* = -\mathcal{R}(P)^\dagger \mathcal{S}(P) + [I - \mathcal{R}(P)^\dagger \mathcal{R}(P)]\Pi, \\
u^* = -\mathcal{R}(P)^\dagger [B^\top \eta + D^\top \zeta + D^\top P\sigma + \rho] + [I - \mathcal{R}(P)^\dagger \mathcal{R}(P)]\nu,
\end{cases} \tag{2.6.13}
$$

where $\Pi \in \mathbb{R}^{m \times n}$ is chosen such that $\Theta^* \in \mathscr{S}[A, C; B, D]$, and $\nu \in \mathcal{U}$ is arbitrary. Furthermore, the value function admits the following representation:

$$V(x) = \langle Px, x \rangle + \mathbb{E}\Big\{2\langle \eta(0), x \rangle + \int_0^\infty \Big[\langle P\sigma, \sigma \rangle + 2\langle \eta, b \rangle + 2\langle \zeta, \sigma \rangle$$
$$- \big|[\mathcal{R}(P)^\dagger]^{\frac{1}{2}}(B^\top \eta + D^\top \zeta + D^\top P\sigma + \rho)\big|^2\Big]dt\Big\}.$$

Proof *Necessity.* Suppose that the pair (Θ^*, u^*) in (2.6.11) is a closed-loop saddle point of Problem $(SDG)_\infty$. Then by Proposition 2.6.4, $(\Theta_1^*, 0; \Theta_2^*, 0)$ is a closed-loop saddle point of Problem $(SDG)_\infty^0$, and hence by Proposition 2.6.6, the ARE (2.6.5) admits a stabilizing solution P. Moreover, from the proof of Proposition 2.6.6, we see that Θ^* is given by (2.6.10) for some matrix $\Pi \in \mathbb{R}^{m \times n}$. To prove (ii), let (X^*, Y^*, Z^*) be the solution of (2.6.4). Then

$$Ru^* + B^\top Y^* + D^\top Z^* + (S + R\Theta^*)X^* + \rho = 0, \quad \text{a.e. a.s.} \qquad (2.6.14)$$

It follows that

$$\begin{aligned}
dY^* &= -\big\{(A+B\Theta^*)^\top Y^* + (C+D\Theta^*)^\top Z^* + \widetilde{Q}X^* + \widetilde{S}^\top u^* + \widetilde{q}\big\}dt + Z^* dW \\
&= -\big\{A^\top Y^* + C^\top Z^* + (Q + S^\top \Theta^*)X^* + S^\top u^* + q \\
&\quad + (\Theta^*)^\top[B^\top Y^* + D^\top Z^* + (S + R\Theta^*)X^* + Ru^* + \rho]\big\}dt + Z^* dW \\
&= -\big\{A^\top Y^* + C^\top Z^* + (Q + S^\top \Theta^*)X^* + S^\top u^* + q\big\}dt + Z^* dW.
\end{aligned}$$

Define for $t \geqslant 0$,

$$\begin{aligned}
\eta(t) &= Y^*(t) - PX^*(t), \\
\zeta(t) &= Z^*(t) - P(C + D\Theta^*)X^*(t) - PDu^*(t) - P\sigma(t).
\end{aligned}$$

Noting that $\mathcal{Q}(P) + \mathcal{S}(P)^\top \Theta^* = 0$, we have

$$\begin{aligned}
d\eta &= dY^* - P dX^* \\
&= -[A^\top Y^* + C^\top Z^* + (Q + S^\top \Theta^*)X^* + S^\top u^* + q]dt + Z^* dW \\
&\quad - P[(A + B\Theta^*)X^* + Bu^* + b]dt - P[(C + D\Theta^*)X^* + Du^* + \sigma]dW \\
&= -\big\{A^\top(\eta + PX^*) + C^\top[\zeta + P(C + D\Theta^*)X^* + PDu^* + P\sigma] \\
&\quad + (Q+S^\top \Theta^*)X^* + S^\top u^* + q + P[(A+B\Theta^*)X^* + Bu^* + b]\big\}dt + \zeta dW \\
&= -\big\{A^\top \eta + C^\top \zeta + \mathcal{Q}(P)X^* + \mathcal{S}(P)^\top \Theta^* X^* + \mathcal{S}(P)^\top u^* \\
&\quad + C^\top P\sigma + Pb + q\big\}dt + \zeta dW \\
&= -\big\{A^\top \eta + C^\top \zeta + \mathcal{S}(P)^\top u^* + C^\top P\sigma + Pb + q\big\}dt + \zeta dW.
\end{aligned}$$

According to (2.6.14), we have (noting that $\mathcal{S}(P) + \mathcal{R}(P)\Theta^* = 0$)

$$0 = B^\top Y^* + D^\top Z^* + (S + R\Theta^*)X^* + Ru^* + \rho$$
$$= B^\top(\eta + PX^*) + D^\top[\zeta + P(C + D\Theta^*)X^* + PDu^* + P\sigma]$$
$$+ (S + R\Theta^*)X^* + Ru^* + \rho$$
$$= [\mathcal{S}(P) + \mathcal{R}(P)\Theta^*]X^* + B^\top\eta + D^\top\zeta + D^\top P\sigma + \rho + \mathcal{R}(P)u^*$$
$$= B^\top\eta + D^\top\zeta + D^\top P\sigma + \rho + \mathcal{R}(P)u^*.$$

Hence,

$$B^\top\eta + D^\top\zeta + D^\top P\sigma + \rho \in \mathscr{R}(\mathcal{R}(P)), \quad \text{a.e. a.s.}$$

and

$$u^* = -\mathcal{R}(P)^\dagger(B^\top\eta + D^\top\zeta + D^\top P\sigma + \rho) + [I - \mathcal{R}(P)^\dagger\mathcal{R}(P)]\nu$$

for some $\nu \in \mathcal{U}$. Consequently,

$$\mathcal{S}(P)^\top u^* = -\mathcal{S}(P)^\top\mathcal{R}(P)^\dagger(B^\top\eta + D^\top\zeta + D^\top P\sigma + \rho)$$
$$+ \mathcal{S}(P)^\top[I - \mathcal{R}(P)^\dagger\mathcal{R}(P)]\nu$$
$$= -\mathcal{S}(P)^\top\mathcal{R}(P)^\dagger(B^\top\eta + D^\top\zeta + D^\top P\sigma + \rho).$$

Then

$$A^\top\eta + C^\top\zeta + \mathcal{S}(P)^\top u^* + C^\top P\sigma + Pb + q$$
$$= A^\top\eta + C^\top\zeta - \mathcal{S}(P)^\top\mathcal{R}(P)^\dagger(B^\top\eta + D^\top\zeta + D^\top P\sigma + \rho)$$
$$+ C^\top P\sigma + Pb + q$$
$$= [A^\top - \mathcal{S}(P)^\top\mathcal{R}(P)^\dagger B^\top]\eta + [C^\top - \mathcal{S}(P)^\top\mathcal{R}(P)^\dagger D^\top]\zeta$$
$$+ [C^\top - \mathcal{S}(P)^\top\mathcal{R}(P)^\dagger D^\top]P\sigma - \mathcal{S}(P)^\top\mathcal{R}(P)^\dagger\rho + Pb + q.$$

Therefore, (η, ζ) is an L^2-stable solution to (2.6.12).

Sufficiency. Let (Θ^*, u^*) be given by (2.6.13), where $\Pi \in \mathbb{R}^{m\times n}$ is chosen so that $\Theta^* \in \mathscr{S}[A, C; B, D]$. Then we have

$$\mathcal{R}(P)\Theta^* + \mathcal{S}(P) = 0, \tag{2.6.15}$$
$$\mathcal{Q}(P) + \mathcal{S}(P)^\top\Theta^* + (\Theta^*)^\top\mathcal{S}(P) + (\Theta^*)^\top\mathcal{R}(P)\Theta^* = 0, \tag{2.6.16}$$
$$B^\top\eta + D^\top\zeta + D^\top P\sigma + \rho = -\mathcal{R}(P)u^*, \tag{2.6.17}$$

and

$$[(\Theta^*)^\top + \mathcal{S}(P)^\top\mathcal{R}(P)^\dagger](B^\top\eta + D^\top\zeta + D^\top P\sigma + \rho)$$
$$= -\Pi^\top[I - \mathcal{R}(P)\mathcal{R}(P)^\dagger]\mathcal{R}(P)u^* = 0. \tag{2.6.18}$$

Take an arbitrary $u \in \mathcal{U}$ and let $X(\cdot) \equiv X(\cdot\,; x, u)$ be the solution to the following closed-loop system:

$$\begin{cases} dX(t) = [(A + B\Theta^*)X(t) + Bu(t) + b(t)]dt \\ \qquad\qquad + [(C + D\Theta^*)X(t) + Du(t) + \sigma(t)]dW(t), \quad t \geqslant 0, \\ X(0) = x. \end{cases}$$

Then

$$\begin{aligned} J(x; \Theta^*X + u) &= \mathbb{E}\int_0^\infty \left[\left\langle \begin{pmatrix} Q & S^\top \\ S & R \end{pmatrix} \begin{pmatrix} X \\ \Theta^*X + u \end{pmatrix}, \begin{pmatrix} X \\ \Theta^*X + u \end{pmatrix} \right\rangle \right. \\ &\qquad\qquad \left. + 2 \left\langle \begin{pmatrix} q \\ \rho \end{pmatrix}, \begin{pmatrix} X \\ \Theta^*X + u \end{pmatrix} \right\rangle \right] dt \\ &= \mathbb{E}\int_0^\infty \Big\{ \langle [Q + S^\top\Theta^* + (\Theta^*)^\top S + (\Theta^*)^\top R\Theta^*]X, X \rangle \\ &\qquad\qquad + 2\langle (S + R\Theta^*)X, u \rangle + \langle Ru, u \rangle \\ &\qquad\qquad + 2\langle q + (\Theta^*)^\top\rho, X \rangle + 2\langle \rho, u \rangle \Big\} dt. \end{aligned} \qquad (2.6.19)$$

Applying Itô's formula to $t \mapsto \langle PX(t), X(t) \rangle$, one has (recalling (2.6.15))

$$\begin{aligned} -\langle Px, x \rangle &= \mathbb{E}\int_0^\infty \Big\{ \langle [P(A + B\Theta^*) + (A + B\Theta^*)^\top P]X, X \rangle \\ &\qquad + \langle P(C + D\Theta^*)X, (C + D\Theta^*)X \rangle + 2\langle PX, Bu + b \rangle \\ &\qquad + 2\langle P(C + D\Theta^*)X, Du + \sigma) \rangle + \langle P(Du + \sigma), Du + \sigma \rangle \Big\} dt \\ &= \mathbb{E}\int_0^\infty \Big\{ \langle [(PA + A^\top P + C^\top PC) + (PB + C^\top PD)\Theta^* \\ &\qquad + (\Theta^*)^\top(B^\top P + D^\top PC) + (\Theta^*)^\top D^\top PD\Theta^*]X, X \rangle \\ &\qquad + 2\langle (B^\top P + D^\top PC + D^\top PD\Theta^*)X, u \rangle + 2\langle P(C + D\Theta^*)X, \sigma \rangle \\ &\qquad + \langle D^\top PDu, u \rangle + 2\langle D^\top P\sigma, u \rangle + 2\langle PX, b \rangle + \langle P\sigma, \sigma \rangle \Big\} dt \\ &= \mathbb{E}\int_0^\infty \Big\{ \langle [Q(P) + S(P)^\top\Theta^* + (\Theta^*)^\top S(P) + (\Theta^*)^\top R(P)\Theta^*]X, X \rangle \\ &\qquad - \langle [Q + S^\top\Theta^* + (\Theta^*)^\top S + (\Theta^*)^\top R\Theta^*]X, X \rangle \\ &\qquad + 2\langle [S(P) + R(P)\Theta^* - (S + R\Theta^*)]X, u \rangle + 2\langle P(C + D\Theta^*)X, \sigma \rangle \\ &\qquad + \langle D^\top PDu, u \rangle + 2\langle D^\top P\sigma, u \rangle + 2\langle PX, b \rangle + \langle P\sigma, \sigma \rangle \Big\} dt \\ &= \mathbb{E}\int_0^\infty \Big\{ 2\langle P(C + D\Theta^*)X, \sigma \rangle + \langle D^\top PDu, u \rangle + 2\langle D^\top P\sigma, u \rangle \\ &\qquad + 2\langle PX, b \rangle + \langle P\sigma, \sigma \rangle - 2\langle (S + R\Theta^*)X, u \rangle \\ &\qquad - \langle [Q + S^\top\Theta^* + (\Theta^*)^\top S + (\Theta^*)^\top R\Theta^*]X, X \rangle \Big\} dt. \end{aligned} \qquad (2.6.20)$$

Applying Itô's formula to $t \mapsto \langle \eta(t), X(t) \rangle$, one has (recalling (2.6.18))

$$
\begin{aligned}
\mathbb{E}\langle \eta(0), x \rangle &= \mathbb{E} \int_0^\infty \Big\{ \big\langle [A - B\mathcal{R}(P)^\dagger \mathcal{S}(P)]^\top \eta + [C - D\mathcal{R}(P)^\dagger \mathcal{S}(P)]^\top \zeta \\
&\quad + [C - D\mathcal{R}(P)^\dagger \mathcal{S}(P)]^\top P\sigma - \mathcal{S}(P)^\top \mathcal{R}(P)^\dagger \rho + Pb + q, \, X \big\rangle \\
&\quad - \langle (A + B\Theta^*)X + Bu + b, \eta \rangle - \langle \zeta, (C + D\Theta^*)X + Du + \sigma \rangle \Big\} dt \\
&= \mathbb{E} \int_0^\infty \Big\{ - \big\langle [\Theta^* + \mathcal{R}(P)^\dagger \mathcal{S}(P)]^\top (B^\top \eta + D^\top \zeta + D^\top P\sigma + \rho), \, X \big\rangle \\
&\quad + \langle P(C + D\Theta^*)X, \sigma \rangle + \big\langle (\Theta^*)^\top \rho + Pb + q, \, X \big\rangle \\
&\quad - \langle Bu + b, \eta \rangle - \langle \zeta, Du + \sigma \rangle \Big\} dt \\
&= \mathbb{E} \int_0^\infty \Big\{ \langle P(C + D\Theta^*)X, \sigma \rangle + \big\langle (\Theta^*)^\top \rho + Pb + q, \, X \big\rangle \\
&\quad - \langle Bu + b, \eta \rangle - \langle \zeta, Du + \sigma \rangle \Big\} dt.
\end{aligned}
\tag{2.6.21}
$$

Combining (2.6.19)–(2.6.21) and recalling (2.6.17), we have

$$
\begin{aligned}
& J(x; \Theta^* X + u) - \langle Px, x \rangle - 2\mathbb{E}\langle \eta(0), x \rangle \\
&= \mathbb{E} \int_0^\infty \Big\{ \langle \mathcal{R}(P)u, u \rangle + 2\langle B^\top \eta + D^\top \zeta + D^\top P\sigma + \rho, u \rangle \\
&\quad + 2\langle b, \eta \rangle + 2\langle \zeta, \sigma \rangle + \langle P\sigma, \sigma \rangle \Big\} dt \\
&= \mathbb{E} \int_0^\infty \Big\{ \langle \mathcal{R}(P)u, u \rangle - 2\langle \mathcal{R}(P)u^*, u \rangle + 2\langle b, \eta \rangle + 2\langle \zeta, \sigma \rangle + \langle P\sigma, \sigma \rangle \Big\} dt \\
&= \mathbb{E} \int_0^\infty \Big\{ \langle \mathcal{R}(P)(u - u^*), u - u^* \rangle - \langle \mathcal{R}(P)u^*, u^* \rangle \\
&\quad + 2\langle b, \eta \rangle + 2\langle \zeta, \sigma \rangle + \langle P\sigma, \sigma \rangle \Big\} dt.
\end{aligned}
$$

Consequently,

$$
\begin{aligned}
& J(x; \Theta_1^* X + u_1, \Theta_2^* X + u_2^*) - J(x; \Theta^* X^* + u^*) \\
&= \mathbb{E} \int_0^\infty \langle \mathcal{R}_1(P)(u_1 - u_1^*), u_1 - u_1^* \rangle dt \geqslant 0,
\end{aligned}
$$

since $\mathcal{R}_1(P) \geqslant 0$. Similarly,

$$
\begin{aligned}
& J(x; \Theta_1^* X + u_1^*, \Theta_2^* X + u_2) - J(x; \Theta^* X^* + u^*) \\
&= \mathbb{E} \int_0^\infty \langle \mathcal{R}_2(P)(u_2 - u_2^*), u_2 - u_2^* \rangle dt \leqslant 0,
\end{aligned}
$$

since $\mathcal{R}_2(P) \leqslant 0$. Therefore, (Θ^*, u^*) is a closed-loop saddle point of Problem $(SQG)_\infty$. Finally, recalling (2.6.17), we have

$$\langle \mathcal{R}(P)u^*, u^* \rangle = \langle \mathcal{R}(P)\mathcal{R}(P)^\dagger \mathcal{R}(P)u^*, u^* \rangle = \langle \mathcal{R}(P)^\dagger \mathcal{R}(P)u^*, \mathcal{R}(P)u^* \rangle$$
$$= \left| [\mathcal{R}(P)^\dagger]^{\frac{1}{2}} (B^\top \eta + D^\top \zeta + D^\top P \sigma + \rho) \right|^2,$$

and hence,

$$V(x) = J(x; \Theta^* X + u^*) = \langle Px, x \rangle + 2\mathbb{E}\langle \eta(0), x \rangle$$
$$+ \mathbb{E} \int_0^\infty \left\{ -\langle \mathcal{R}(P)u^*, u^* \rangle + 2\langle b, \eta \rangle + 2\langle \zeta, \sigma \rangle + \langle P\sigma, \sigma \rangle \right\} dt$$
$$= \langle Px, x \rangle + 2\mathbb{E}\langle \eta(0), x \rangle + \mathbb{E} \int_0^\infty \left[\langle P\sigma, \sigma \rangle + 2\langle \eta, b \rangle + 2\langle \zeta, \sigma \rangle \right.$$
$$\left. - \left| [\mathcal{R}(P)^\dagger]^{\frac{1}{2}} (B^\top \eta + D^\top \zeta + D^\top P \sigma + \rho) \right|^2 \right] dt.$$

This completes the proof.   □

To conclude this section, we present some examples illustrating how the "stabilizing solution" of algebraic Riccati equations plays an important role in the study of closed-loop saddle points.

The first example shows that the algebraic Riccati equation may only admits non-stabilizing solutions even if the system (2.6.3) is stabilizable.

Example 2.6.8 Consider the one-dimensional state equation

$$\begin{cases} dX(t) = -\dfrac{1}{2}X(t)dt + [u_1(t) + u_2(t)]dW(t), \quad t \geqslant 0, \\ X(0) = x, \end{cases}$$

and the performance functional

$$J(x; u_1, u_2) = \mathbb{E} \int_0^\infty \left\langle \begin{pmatrix} 1 & 1 & -1 \\ 1 & 1 & 0 \\ -1 & 0 & -1 \end{pmatrix} \begin{pmatrix} X(t) \\ u_1(t) \\ u_2(t) \end{pmatrix}, \begin{pmatrix} X(t) \\ u_1(t) \\ u_2(t) \end{pmatrix} \right\rangle dt.$$

In this example,

$$A = -\frac{1}{2}, \quad B = (0, 0), \quad C = 0, \quad D = (1, 1),$$
$$Q = 1, \quad S = \begin{pmatrix} 1 \\ -1 \end{pmatrix}, \quad R = \begin{pmatrix} 1 & 0 \\ 0 & -1 \end{pmatrix}.$$

It is clear that $\Theta = (\Theta_1, \Theta_2)^\top \in \mathscr{S}[A, C; B, D]$ if and only if

$$-1 + (\Theta_1 + \Theta_2)^2 = 2(A + B\Theta) + (C + D\Theta)^2 < 0,$$

or equivalently, if and only if

$$-1 < \Theta_1 + \Theta_2 < 1.$$

Note that $R + D^\top PD$ is invertible for all $P \in \mathbb{R}$ with inverse

$$(R + D^\top PD)^{-1} = \begin{pmatrix} P+1 & P \\ P & P-1 \end{pmatrix}^{-1} = \begin{pmatrix} -P+1 & P \\ P & -P-1 \end{pmatrix}.$$

Then the corresponding ARE reads

$$\begin{aligned} 0 &= PA + A^\top P + C^\top PC + Q \\ &\quad - (PB + C^\top PD + S^\top)(R + D^\top PD)^\dagger(B^\top P + D^\top PC + S) \\ &= -P + 1 - (1, -1)\begin{pmatrix} -P+1 & P \\ P & -P-1 \end{pmatrix}\begin{pmatrix} 1 \\ -1 \end{pmatrix} \\ &= 3P + 1. \end{aligned}$$

Thus, $P = -1/3$ and

$$R_{11} + D_1^\top PD_1 = \frac{2}{3} \geqslant 0, \quad R_{22} + D_2^\top PD_2 = -\frac{4}{3} \leqslant 0.$$

Also, the range condition

$$\mathscr{R}(B^\top P + D^\top PC + S) \subseteq \mathscr{R}(R + D^\top PD)$$

holds automatically since $R + D^\top PD$ is invertible. However, for any $\Pi \in \mathbb{R}$,

$$[I - \mathcal{R}(P)^\dagger \mathcal{R}(P)]\Pi - \mathcal{R}(P)^\dagger S(P) = \left(-\frac{5}{3}, -\frac{1}{3}\right)^\top,$$

which is not a stabilizer of the system (2.6.3). Hence, by Theorem 2.6.7, the above problem does not admit closed-loop saddle points.

Next we give an example of Problem (SDG)$_\infty^0$ which admits uncountably many closed-loop saddle points. This example also tells us when the algebraic Riccati equation is solvable, $-\mathcal{R}(P)^\dagger S(P)$ might not be a stabilizer of the system (2.6.3) in general, and we should carefully choose Π so that $[I - \mathcal{R}(P)^\dagger \mathcal{R}(P)]\Pi - \mathcal{R}(P)^\dagger S(P)$ is a closed-loop saddle point of the game.

Example 2.6.9 Consider the one-dimensional state equation

$$
\begin{cases}
dX(t) = -\left[\dfrac{1}{4}X(t) + \dfrac{1}{2}u_2(t)\right]dt + [-X(t) + u_1(t)]dW(t), & t \geqslant 0, \\
X(0) = x,
\end{cases}
$$

and the performance functional

$$
J(x; u_1, u_2) = \mathbb{E}\int_0^\infty \left\langle \begin{pmatrix} \frac{1}{2} & -1 & -\frac{1}{2} \\ -1 & 1 & 0 \\ -\frac{1}{2} & 0 & 0 \end{pmatrix} \begin{pmatrix} X(t) \\ u_1(t) \\ u_2(t) \end{pmatrix}, \begin{pmatrix} X(t) \\ u_1(t) \\ u_2(t) \end{pmatrix} \right\rangle dt.
$$

In this example,

$$
A = -\frac{1}{4}, \quad B = \left(0, -\frac{1}{2}\right), \quad C = -1, \quad D = (1, 0),
$$

$$
Q = \frac{1}{2}, \quad S = \begin{pmatrix} -1 \\ -\frac{1}{2} \end{pmatrix}, \quad R = \begin{pmatrix} 1 & 0 \\ 0 & 0 \end{pmatrix}.
$$

Clearly, $\Theta = (\Theta_1, \Theta_2)^{\mathsf{T}} \in \mathscr{S}[A, C; B, D]$ if and only if

$$
2\left(-\frac{1}{4} - \frac{1}{2}\Theta_2\right) + (-1 + \Theta_1)^2 = 2(A + B\Theta) + (C + D\Theta)^2 < 0,
$$

or equivalently, if and only if

$$
\Theta_1^2 - 2\Theta_1 + \frac{1}{2} < \Theta_2.
$$

The corresponding ARE reads

$$
\begin{aligned}
0 &= PA + A^{\mathsf{T}}P + C^{\mathsf{T}}PC + Q \\
&\quad - (PB + C^{\mathsf{T}}PD + S^{\mathsf{T}})(R + D^{\mathsf{T}}PD)^{\dagger}(B^{\mathsf{T}}P + D^{\mathsf{T}}PC + S) \\
&= \frac{1}{2}(P + 1) - \left(-(P + 1), -\frac{1}{2}(P + 1)\right)\begin{pmatrix} P + 1 & 0 \\ 0 & 0 \end{pmatrix}^{\dagger}\begin{pmatrix} -(P + 1) \\ -\frac{1}{2}(P + 1) \end{pmatrix} \\
&= \frac{1}{2}(P + 1) - \frac{1}{4}(P + 1)^2(2, 1)\begin{pmatrix} P + 1 & 0 \\ 0 & 0 \end{pmatrix}^{\dagger}\begin{pmatrix} 2 \\ 1 \end{pmatrix}.
\end{aligned} \tag{2.6.22}
$$

It is easy to verify that $P = -1$ is the unique solution of (2.6.22). Thus,

$$
\mathcal{R}(P) = R + D^{\mathsf{T}}PD = \begin{pmatrix} 0 & 0 \\ 0 & 0 \end{pmatrix}, \quad \mathcal{S}(P) = B^{\mathsf{T}}P + D^{\mathsf{T}}PC + S = \begin{pmatrix} 0 \\ 0 \end{pmatrix}.
$$

Hence, all the conditions

$$R_{11} + D_1^\top P D_1 \geqslant 0, \quad R_{22} + D_2^\top P D_2 \leqslant 0,$$
$$\mathscr{R}(B^\top P + D^\top P C + S) \subseteq \mathscr{R}(R + D^\top P D)$$

hold. By Theorem 2.6.7, we see that

$$(\Theta_1, \nu_1; \Theta_2, \nu_2) \quad \text{with } \Theta_1^2 - 2\Theta_1 + \frac{1}{2} < \Theta_2, \ \nu_1, \nu_2 \in L_\mathbb{F}^2(\mathbb{R})$$

are all the closed-loop saddle points of the above problem. However,

$$-\mathcal{R}(P)^\dagger S(P) = (0, 0)^\top \notin \mathscr{S}[A, C; B, D].$$

From this example, we also see that even if $-\mathcal{R}(P)^\dagger S(P)$ is not a stabilizer of the system, Problem $(SDG)_\infty$ may still admit closed-loop saddle points, thanks to the fact that we can properly choose $\Pi \neq 0$ so that the term $[I - \mathcal{R}(P)^\dagger \mathcal{R}(P)]\Pi$ could play a role.

Finally, we present an example showing that not all of the stabilizers are necessarily closed-loop saddle points of the game. It may happens that the system (2.6.3) has more than one (uncountably many) stabilizer, while the closed-loop saddle point is unique.

Example 2.6.10 Consider the one-dimensional state equation

$$\begin{cases} dX(t) = [-8X(t) + u_1(t) - u_2(t)]dt + [u_1(t) + u_2(t)]dW(t), \quad t \geqslant 0, \\ X(0) = x, \end{cases}$$

and the performance functional

$$J(x; u_1, u_2) = \mathbb{E} \int_0^\infty \left[12|X(t)|^2 + |u_1(t)|^2 - |u_2(t)|^2 \right] dt.$$

In this example,

$$A = -8, \quad B = (1, -1), \quad C = 0, \quad D = (1, 1),$$
$$Q = 12, \quad S = \begin{pmatrix} 0 \\ 0 \end{pmatrix}, \quad R = \begin{pmatrix} 1 & 0 \\ 0 & -1 \end{pmatrix}.$$

Again, $\Theta = (\Theta_1, \Theta_2)^\top \in \mathscr{S}[A, C; B, D]$ if and only if

$$- 16 + 2(\Theta_1 - \Theta_2) + (\Theta_1 + \Theta_2)^2 < 0. \tag{2.6.23}$$

The corresponding ARE reads

$$0 = PA + A^\top P + C^\top PC + Q$$
$$- (PB + C^\top PD + S^\top)(R + D^\top PD)^\dagger(B^\top P + D^\top PC + S)$$
$$= -16P + 12 - P^2(1, -1)\begin{pmatrix} P+1 & P \\ P & P-1 \end{pmatrix}^\dagger \begin{pmatrix} 1 \\ -1 \end{pmatrix}$$
$$= -16P + 12 - P^2(1, -1)\begin{pmatrix} -P+1 & P \\ P & -P-1 \end{pmatrix}\begin{pmatrix} 1 \\ -1 \end{pmatrix}$$
$$= 4P^3 - 16P + 12,$$

which has three solutions:

$$P_1 = 1, \quad P_2 = \frac{-1 + \sqrt{13}}{2}, \quad P_3 = \frac{-1 - \sqrt{13}}{2}.$$

All of them satisfy the range condition

$$\mathscr{R}(B^\top P + D^\top PC + S) \subseteq \mathscr{R}(R + D^\top PD),$$

since $R + D^\top PD$ is invertible for any $P \in \mathbb{R}$. However, only $P_1 = 1$ satisfies

$$R_{11} + D_1^\top PD_1 \geqslant 0, \quad R_{22} + D_2^\top PD_2 \leqslant 0.$$

For any $\Pi \in \mathbb{R}$,

$$[I - \mathcal{R}(P_1)^\dagger \mathcal{R}(P_1)]\Pi - \mathcal{R}(P_1)^\dagger \mathcal{S}(P_1) = (1, -3)^\top,$$

which satisfies (2.6.23) and hence is a stabilizer of the system (2.6.3). By Theorem 2.6.7, the above problem admits a unique closed-loop saddle point $(1, 0; -3, 0)$.

On the other hand, by verifying (2.6.23), we see that all the following

$$(\Theta_1^*, \Theta_2^*) = (1, -3), \qquad (\Theta_1, \Theta_2) = (0, 0),$$
$$(\Theta_1^*, \Theta_2) = (1, 0), \qquad (\Theta_1, \Theta_2^*) = (0, -3),$$

are stabilizers of the system (2.6.3), but only (Θ_1^*, Θ_2^*) is the closed-loop saddle point of the problem.

Chapter 3
Mean-Field Linear-Quadratic Optimal Controls

Abstract This chapter is concerned with a more general class of linear-quadratic optimal control problems, the mean-field linear-quadratic optimal control problem, in which the expectations of the state process and the control are involved. Two differential Riccati equations are introduced for the problem. The strongly regular solvability of these two Riccati equations is proved to be equivalent to the uniform convexity of the cost functional. In terms of the solutions to the Riccati equations, the unique optimal control is obtained as a linear feedback of the state process and its expectation. An application of the mean-field linear-quadratic optimal control theory is presented, in which analytical optimal portfolio policies are constructed for a continuous-time mean-variance portfolio selection problem. The mean-field linear-quadratic optimal control problem over an infinite horizon is also studied.

Keywords Mean-field · Linear-quadratic · Optimal control · Riccati equation · Open-loop solvability · Uniform convexity · Mean-variance portfolio selection

In this chapter, we study a more general class of linear-quadratic optimal control problems, in which the expectations of the state process and the control, called the *mean-field*, are involved. As before, we shall have a complete probability space $(\Omega, \mathcal{F}, \mathbb{P})$ on which a standard one-dimensional Brownian motion $W = \{W(t); 0 \leqslant t < \infty\}$ is defined. We denote by $\mathbb{F} = \{\mathcal{F}_t\}_{t \geqslant 0}$ the usual augmentation of the natural filtration generated by W and employ throughout this chapter the notation of Chap. 1.

3.1   Problem Formulation and General Considerations

Consider the following controlled state equation on a finite horizon $[t, T]$:

$$
\begin{cases}
dX(s) = \big\{ A(s)X(s) + \bar{A}(s)\mathbb{E}[X(s)] \\
\qquad\quad + B(s)u(s) + \bar{B}(s)\mathbb{E}[u(s)] + b(s) \big\} ds \\
\qquad\quad + \big\{ C(s)X(s) + \bar{C}(s)\mathbb{E}[X(s)] \\
\qquad\quad + D(s)u(s) + \bar{D}(s)\mathbb{E}[u(s)] + \sigma(s) \big\} dW(s), \\
X(t) = \xi.
\end{cases}
\tag{3.1.1}
$$

The *initial pair* (t, ξ) belongs to

$$
\mathscr{D} = \big\{ (t, \xi) \mid t \in [0, T], \ \xi \in \mathcal{X}_t \equiv L^2_{\mathcal{F}_t}(\Omega; \mathbb{R}^n) \big\},
$$

and the control u is taken from $\mathcal{U}[t, T] \equiv L^2_{\mathbb{F}}(t, T; \mathbb{R}^m)$, which is the same as Problem (SLQ). Note that in (3.1.1), the expectations of $X(s)$ and $u(s)$ are involved. We therefore call (3.1.1) a *mean-field* SDE. The cost functional we are considering is a quadratic form and also involves the expectations of $X(s)$ and $u(s)$:

$$
J(t, \xi; u) \triangleq L(t, \xi; u) + \bar{L}(t, \xi; u),
\tag{3.1.2}
$$

where

$$
\begin{aligned}
L(t, x; u) = \mathbb{E}\bigg\{ & \langle GX(T), X(T) \rangle + 2\langle g, X(T) \rangle \\
& + \int_t^T \bigg[\bigg\langle \begin{pmatrix} Q(s) & S(s)^\top \\ S(s) & R(s) \end{pmatrix} \begin{pmatrix} X(s) \\ u(s) \end{pmatrix}, \begin{pmatrix} X(s) \\ u(s) \end{pmatrix} \bigg\rangle \\
& + 2 \bigg\langle \begin{pmatrix} q(s) \\ \rho(s) \end{pmatrix}, \begin{pmatrix} X(s) \\ u(s) \end{pmatrix} \bigg\rangle \bigg] ds \bigg\}, \\
\bar{L}(t, \xi; u) = & \langle \bar{G}\mathbb{E}[X(T)], \mathbb{E}[X(T)] \rangle + 2\langle \bar{g}, \mathbb{E}[X(T)] \rangle \\
& + \int_t^T \bigg[\bigg\langle \begin{pmatrix} \bar{Q}(s) & \bar{S}(s)^\top \\ \bar{S}(s) & \bar{R}(s) \end{pmatrix} \begin{pmatrix} \mathbb{E}[X(s)] \\ \mathbb{E}[u(s)] \end{pmatrix}, \begin{pmatrix} \mathbb{E}[X(s)] \\ \mathbb{E}[u(s)] \end{pmatrix} \bigg\rangle \\
& + 2 \bigg\langle \begin{pmatrix} \bar{q}(s) \\ \bar{\rho}(s) \end{pmatrix}, \begin{pmatrix} \mathbb{E}[X(s)] \\ \mathbb{E}[u(s)] \end{pmatrix} \bigg\rangle \bigg] ds.
\end{aligned}
$$

To make a precise statement of the *mean-field linear-quadratic optimal control problem* (MFLQ problem, for short), we introduce the following assumptions, which are comparable with (H1) and (H2) introduced in Chap. 1.

(A1) The coefficients of (3.1.1) satisfy

$$\begin{cases} A, \bar{A} \in L^1(0, T; \mathbb{R}^{n \times n}), & B, \bar{B} \in L^2(0, T; \mathbb{R}^{n \times m}), \\ C, \bar{C} \in L^2(0, T; \mathbb{R}^{n \times n}), & D, \bar{D} \in L^\infty(0, T; \mathbb{R}^{n \times m}), \\ b \in L_{\mathbb{F}}^2(\Omega; L^1(0, T; \mathbb{R}^n)), & \sigma \in L_{\mathbb{F}}^2(0, T; \mathbb{R}^n). \end{cases}$$

(A2) The weighting matrices in (3.1.2) satisfy

$$\begin{cases} Q, \bar{Q} \in L^1(0, T; \mathbb{S}^n), & S, \bar{S} \in L^2(0, T; \mathbb{R}^{m \times n}), \\ R, \bar{R} \in L^\infty(0, T; \mathbb{S}^m), & G, \bar{G} \in \mathbb{S}^n, \\ q \in L_{\mathbb{F}}^2(\Omega; L^1(0, T; \mathbb{R}^n)), & \bar{q} \in L^1(0, T; \mathbb{R}^n), \\ \rho \in L_{\mathbb{F}}^2(0, T; \mathbb{R}^m), & \bar{\rho} \in L^2(0, T; \mathbb{R}^m), \\ g \in L_{\mathcal{F}_T}^2(\Omega; \mathbb{R}^n), & \bar{g} \in \mathbb{R}^n. \end{cases}$$

It can be shown that under the assumptions (A1) and (A2), for any initial pair $(t, \xi) \in \mathscr{D}$ and control $u \in \mathcal{U}[t, T]$, the state Equation (3.1.1) has a unique solution $X(\cdot) \equiv X(\cdot; t, \xi, u) \in \mathcal{X}[t, T] \equiv L_{\mathbb{F}}^2(\Omega; C([t, T]; \mathbb{R}^n))$, and the cost functional $J(t, \xi; u)$ is therefore well-defined. The MFLQ problem is then stated as follows.

Problem (MFLQ). For given initial pair $(t, \xi) \in \mathscr{D}$, find a control $u^* \in \mathcal{U}[t, T]$ such that

$$J(t, \xi; u^*) = \inf_{u \in \mathcal{U}[t,T]} J(t, \xi; u) \equiv V(t, \xi). \qquad (3.1.3)$$

An element $u^* \in \mathcal{U}[t, T]$ that satisfies (3.1.3) is called an *open-loop optimal control* of Problem (MFLQ) for the initial pair (t, ξ). The solution $X^*(\cdot) \equiv X(\cdot; t, \xi, u^*)$ of (3.1.1) corresponding to an optimal control u^* is called an *optimal state process*. The function $V(t, \xi)$ is called the *value function* of Problem (MFLQ). We shall denote by Problem (MFLQ)^0 the particular case of Problem (MFLQ) where the non-homogeneous terms b, σ of the state equation and the coefficients of the linear terms $g, \bar{g}, q, \bar{q}, \rho, \bar{\rho}$ in the cost functional are zero. The cost functional and value function of Problem (MFLQ)^0 will be denoted by $J^0(t, \xi; u)$ and $V^0(t, \xi)$, respectively.

Definition 3.1.1 Problem (MFLQ) is said to be

(i) *finite at the initial pair* $(t, \xi) \in \mathscr{D}$ if $V(t, \xi) > -\infty$;
(ii) *finite at* $t \in [0, T]$ if $V(t, \xi) > -\infty$ for all $\xi \in \mathcal{X}_t$;
(iii) *finite* if it is finite at all $t \in [0, T]$.

Definition 3.1.2 Problem (MFLQ) is said to be

(i) *(uniquely) open-loop solvable at the initial pair* $(t, \xi) \in \mathscr{D}$ if there exists a (unique) $u^* \in \mathcal{U}[t, T]$ such that (3.1.3) holds;
(ii) *(uniquely) open-loop solvable at* t if for any $\xi \in \mathcal{X}_t$, there exists a (unique) $u^* \in \mathcal{U}[t, T]$ such that (3.1.3) holds;

(iii) *(uniquely) open-loop solvable* if it is (uniquely) open-loop solvable at all $t \in [0, T)$.

It should be noted that Problem (MFLQ) might be finite but not solvable at an initial pair.

3.2 Open-Loop Solvability and Mean-Field FBSDEs

In order to study the open-loop solvability of Problem (MFLQ), we introduce the following mean-field BSDE:

$$
\begin{cases}
dY(s) = -\big\{ A^{\mathsf{T}} Y + \bar{A}^{\mathsf{T}} \mathbb{E}[Y] + C^{\mathsf{T}} Z + \bar{C}^{\mathsf{T}} \mathbb{E}[Z] + QX + \bar{Q} \mathbb{E}[X] \\
\qquad\qquad + S^{\mathsf{T}} u + \bar{S}^{\mathsf{T}} \mathbb{E}[u] + q + \bar{q} \big\} ds + Z dW, \quad s \in [t, T], \qquad (3.2.1) \\
Y(T) = GX(T) + \bar{G} \mathbb{E}[X(T)] + g + \bar{g},
\end{cases}
$$

where X is the solution to the state Equation (3.1.1) corresponding to some initial pair $(t, \xi) \in \mathscr{D}$ and control $u \in \mathcal{U}[t, T]$. This mean-field BSDE is called the *adjoint equation* associated with (3.1.1). Similar to the standard BSDE theory, the mean-field BSDE (3.2.1) admits a unique adapted solution

$$
(Y, Z) \in L^2_{\mathbb{F}}(\Omega; C([t, T]; \mathbb{R}^n)) \times L^2_{\mathbb{F}}(t, T; \mathbb{R}^n).
$$

Proposition 3.2.1 *Let (A1)–(A2) hold, and let $(t, \xi) \in \mathscr{D}$ be a given initial pair. The following holds for any scalar $\lambda \in \mathbb{R}$ and controls $u, v \in \mathcal{U}[t, T]$:*

$$
\begin{aligned}
& J(t, \xi; u + \lambda v) - J(t, \xi; u) \\
& = \lambda^2 J^0(t, 0; v) + 2\lambda \mathbb{E} \int_t^T \big\langle B^{\mathsf{T}} Y + \bar{B}^{\mathsf{T}} \mathbb{E}[Y] + D^{\mathsf{T}} Z + \bar{D}^{\mathsf{T}} \mathbb{E}[Z] \\
& \quad + SX + \bar{S} \mathbb{E}[X] + Ru + \bar{R} \mathbb{E}[u] + \rho + \bar{\rho}, v \big\rangle ds, \qquad\qquad (3.2.2)
\end{aligned}
$$

where X is the state process corresponding to (t, ξ, u), and (Y, Z) is the adapted solution to the associated adjoint equation (3.2.1). Consequently, the map $u \mapsto J(t, \xi; u)$ is Fréchet differentiable with the Fréchet derivative given by

$$
\begin{aligned}
\mathcal{D}_u J(t, \xi; u) = 2\big\{ & B^{\mathsf{T}} Y + \bar{B}^{\mathsf{T}} \mathbb{E}[Y] + D^{\mathsf{T}} Z + \bar{D}^{\mathsf{T}} \mathbb{E}[Z] \\
& + SX + \bar{S} \mathbb{E}[X] + Ru + \bar{R} \mathbb{E}[u] + \rho + \bar{\rho} \big\}.
\end{aligned}
$$

Proof Let $X^\lambda(\cdot) = X(\cdot; t, \xi, u + \lambda v)$ be the state process corresponding to the control $u + \lambda v$ and with initial pair (t, ξ), and let $\mathring{X}_0^v$ be the solution to the mean-field SDE

$$\begin{cases} d\mathring{X}_0^v(s) = \{A\mathring{X}_0^v + \bar{A}\mathbb{E}[\mathring{X}_0^v] + Bv + \bar{B}\mathbb{E}[v]\}ds \\ \qquad\qquad + \{C\mathring{X}_0^v + \bar{C}\mathbb{E}[\mathring{X}_0^v] + Dv + \bar{D}\mathbb{E}[v]\}dW, \quad s \in [t, T], \\ \mathring{X}_0^v(t) = 0. \end{cases}$$

It is easily seen, by the linearity of the state equation, that $X^\lambda = X + \lambda\mathring{X}_0^v$. Substituting this relation into the expression of $J(t, \xi; u + \lambda v)$, we obtain by a direct computation that

$$\begin{aligned} J(t, \xi&; u + \lambda v) - J(t, \xi; u) \\ &= \lambda^2 J^0(t, 0; v) + 2\lambda\mathbb{E}\Big\{\langle GX(T) + \bar{G}\mathbb{E}[X(T)] + g + \bar{g}, \mathring{X}_0^v(T)\rangle \\ &\quad + \int_t^T \Big[\langle QX + \bar{Q}\mathbb{E}[X] + S^\top u + \bar{S}^\top\mathbb{E}[u] + q + \bar{q}, \mathring{X}_0^v\rangle \\ &\quad + \langle SX + \bar{S}\mathbb{E}[X] + Ru + \bar{R}\mathbb{E}[u] + \rho + \bar{\rho}, v\rangle\Big]ds\Big\}. \end{aligned} \qquad (3.2.3)$$

Now by applying Itô's formula to $s \mapsto \langle Y(s), \mathring{X}_0^v(s)\rangle$, we have

$$\begin{aligned} \mathbb{E}\langle GX&(T) + \bar{G}\mathbb{E}[X(T)] + g + \bar{g}, \mathring{X}_0(T)\rangle \\ &= \mathbb{E}\int_t^T \Big\{ -\langle A^\top Y + \bar{A}^\top\mathbb{E}[Y] + C^\top Z + \bar{C}^\top\mathbb{E}[Z] + QX + \bar{Q}\mathbb{E}[X] \\ &\qquad\qquad + S^\top u + \bar{S}^\top\mathbb{E}[u] + q + \bar{q}, \mathring{X}_0^v\rangle \\ &\qquad + \langle A\mathring{X}_0^v + \bar{A}\mathbb{E}[\mathring{X}_0^v] + Bv + \bar{B}\mathbb{E}[v], Y\rangle \\ &\qquad + \langle C\mathring{X}_0^v + \bar{C}\mathbb{E}[\mathring{X}_0^v] + Dv + \bar{D}\mathbb{E}[v], Z\rangle\Big\}ds \\ &= \mathbb{E}\int_t^T \Big\{\langle B^\top Y + \bar{B}^\top\mathbb{E}[Y] + D^\top Z + \bar{D}^\top\mathbb{E}[Z], v\rangle \\ &\qquad - \langle QX + \bar{Q}\mathbb{E}[X] + S^\top u + \bar{S}^\top\mathbb{E}[u] + q + \bar{q}, \mathring{X}_0^v\rangle\Big\}ds. \end{aligned} \qquad (3.2.4)$$

Substituting (3.2.4) into (3.2.3) yields (3.2.2). □

Theorem 3.2.2 *Let (A1)–(A2) hold, and let $(t, \xi) \in \mathcal{D}$ be given. Let $u \in \mathcal{U}[t, T]$ and (X, Y, Z) be the adapted solution to the following (decoupled) mean-field FBSDE:*

$$\begin{cases} dX(s) = \{AX + \bar{A}\mathbb{E}[X] + Bu + \bar{B}\mathbb{E}[u] + b\}ds \\ \qquad\qquad + \{CX + \bar{C}\mathbb{E}[X] + Du + \bar{D}\mathbb{E}[u] + \sigma\}dW, \\ dY(s) = -\{A^\top Y + \bar{A}^\top\mathbb{E}[Y] + C^\top Z + \bar{C}^\top\mathbb{E}[Z] + QX + \bar{Q}\mathbb{E}[X] \\ \qquad\qquad + S^\top u + \bar{S}^\top\mathbb{E}[u] + q + \bar{q}\}ds + ZdW, \\ X(t) = \xi, \quad Y(T) = GX(T) + \bar{G}\mathbb{E}[X(T)] + g + \bar{g}. \end{cases} \qquad (3.2.5)$$

Then u is an open-loop optimal control of Problem (MFLQ) for the initial pair (t, ξ)
if and only if

$$J^0(t, 0; v) \geqslant 0, \quad \forall v \in \mathcal{U}[t, T], \tag{3.2.6}$$

and the following holds almost everywhere on $[t, T]$ *and almost surely:*

$$B^\top Y + D^\top Z + SX + Ru + \rho$$
$$+ \bar{B}^\top \mathbb{E}[Y] + \bar{D}^\top \mathbb{E}[Z] + \bar{S}\mathbb{E}[X] + \bar{R}\mathbb{E}[u] + \bar{\rho} = 0. \tag{3.2.7}$$

Proof By (3.2.2), we see that u is an optimal control of Problem (MFLQ) for the
initial pair (t, ξ) if and only if

$$\lambda^2 J^0(t, 0; v) + \lambda \mathbb{E} \int_t^T \langle \mathcal{D}_u J(t, \xi; u)(s), v(s) \rangle ds$$
$$= J(t, \xi; u + \lambda v) - J(t, \xi; u) \geqslant 0, \quad \forall \lambda \in \mathbb{R}, \ \forall v \in \mathcal{U}[t, T],$$

which is equivalent to (3.2.6), together with the following:

$$\mathbb{E} \int_t^T \langle \mathcal{D}_u J(t, \xi; u)(s), v(s) \rangle ds = 0, \quad \forall v \in \mathcal{U}[t, T].$$

The result follows immediately.   □

3.3 A Hilbert Space Point of View

According to Proposition 3.2.1, the cost functional $J(t, \xi; u)$ can be written as

$$J(t, \xi; u) = J^0(t, 0; u) + \langle \mathcal{D}_u J(t, \xi; 0), u \rangle + J(t, \xi; 0).$$

For fixed initial pair $(t, \xi) \in \mathcal{D}$, the second term is linear in the control variable u,
and the last term is a constant. We now look at the structure of $J^0(t, 0; u)$ from a
Hilbert space point of view.

For an initial pair $(t, \xi) \in \mathcal{D}$ and a control $u \in \mathcal{U}[t, T]$, we denote by $\mathring{X}_\xi^u$ the
solution to the homogeneous state equation

$$\begin{cases} d\mathring{X}_\xi^u(s) = \left\{ A\mathring{X}_\xi^u + \bar{A}\mathbb{E}[\mathring{X}_\xi^u] + Bu + \bar{B}\mathbb{E}[u] \right\} ds \\ \qquad\qquad + \left\{ C\mathring{X}_\xi^u + \bar{C}\mathbb{E}[\mathring{X}_\xi^u] + Du + \bar{D}\mathbb{E}[u] \right\} dW, \tag{3.3.1} \\ \mathring{X}_\xi^u(t) = \xi. \end{cases}$$

In terms of $\mathring{X}_\xi^u$, the cost functional $J^0(t, \xi; u)$ becomes

$$J^0(t, \xi; u) = \mathbb{E}\Bigg\{\langle G\mathring{X}_\xi^u(T), \mathring{X}_\xi^u(T)\rangle + \langle \bar{G}\mathbb{E}[\mathring{X}_\xi^u(T)], \mathbb{E}[\mathring{X}_\xi^u(T)]\rangle$$
$$+ \int_t^T \left\langle \begin{pmatrix} Q(s) & S(s)^\top \\ S(s) & R(s) \end{pmatrix} \begin{pmatrix} \mathring{X}_\xi^u(s) \\ u(s) \end{pmatrix}, \begin{pmatrix} \mathring{X}_\xi^u(s) \\ u(s) \end{pmatrix} \right\rangle ds$$
$$+ \int_t^T \left\langle \begin{pmatrix} \bar{Q}(s) & \bar{S}(s)^\top \\ \bar{S}(s) & \bar{R}(s) \end{pmatrix} \begin{pmatrix} \mathbb{E}[\mathring{X}_\xi^u(s)] \\ \mathbb{E}[u(s)] \end{pmatrix}, \begin{pmatrix} \mathbb{E}[\mathring{X}_\xi^u(s)] \\ \mathbb{E}[u(s)] \end{pmatrix} \right\rangle ds\Bigg\}.$$

When $\xi = 0$, we can define two bounded linear operators

$$\mathcal{L}_t : \mathcal{U}[t, T] \to \mathcal{X}[t, T] \quad \text{and} \quad \widehat{\mathcal{L}}_t : \mathcal{U}[t, T] \to \mathcal{X}_T$$

by

$$\mathcal{L}_t u = \mathring{X}_0^u \quad \text{and} \quad \widehat{\mathcal{L}}_t u = \mathring{X}_0^u(T),$$

respectively. Denote by $\mathcal{L}_t^*$, $\widehat{\mathcal{L}}_t^*$, and $\mathbb{E}^*$ the adjoints of $\mathcal{L}_t$, $\widehat{\mathcal{L}}_t$, and the expectation operator $\mathbb{E}$, respectively. Then

$$J^0(t, 0; u) = \mathbb{E}\Bigg\{\langle G\mathring{X}_0^u(T), \mathring{X}_0^u(T)\rangle + \langle \bar{G}\mathbb{E}[\mathring{X}_0^u(T)], \mathbb{E}[\mathring{X}_0^u(T)]\rangle$$
$$+ \int_t^T \left\langle \begin{pmatrix} Q(s) & S(s)^\top \\ S(s) & R(s) \end{pmatrix} \begin{pmatrix} \mathring{X}_0^u(s) \\ u(s) \end{pmatrix}, \begin{pmatrix} \mathring{X}_0^u(s) \\ u(s) \end{pmatrix} \right\rangle ds$$
$$+ \int_t^T \left\langle \begin{pmatrix} \bar{Q}(s) & \bar{S}(s)^\top \\ \bar{S}(s) & \bar{R}(s) \end{pmatrix} \begin{pmatrix} \mathbb{E}[\mathring{X}_0^u(s)] \\ \mathbb{E}[u(s)] \end{pmatrix}, \begin{pmatrix} \mathbb{E}[\mathring{X}_0^u(s)] \\ \mathbb{E}[u(s)] \end{pmatrix} \right\rangle ds\Bigg\}$$
$$= \langle G\widehat{\mathcal{L}}_t u, \widehat{\mathcal{L}}_t u\rangle + \langle \bar{G}\mathbb{E}[\widehat{\mathcal{L}}_t u], \mathbb{E}[\widehat{\mathcal{L}}_t u]\rangle$$
$$+ \langle Q\mathcal{L}_t u, \mathcal{L}_t u\rangle + 2\langle S\mathcal{L}_t u, u\rangle + \langle Ru, u\rangle$$
$$+ \langle \bar{Q}\mathbb{E}[\mathcal{L}_t u], \mathbb{E}[\mathcal{L}_t u]\rangle + 2\langle \bar{S}\mathbb{E}[\mathcal{L}_t u], \mathbb{E}[u]\rangle + \langle \bar{R}\mathbb{E}[u], \mathbb{E}[u]\rangle.$$

Let

$$\mathcal{M}_t = \widehat{\mathcal{L}}_t^*(G + \mathbb{E}^* \bar{G}\mathbb{E})\widehat{\mathcal{L}}_t + \mathcal{L}_t^*(Q + \mathbb{E}^* \bar{Q}\mathbb{E})\mathcal{L}_t$$
$$+ (S + \mathbb{E}^* \bar{S}\mathbb{E})\mathcal{L}_t + \mathcal{L}_t^*(S^\top + \mathbb{E}^* \bar{S}^\top \mathbb{E}) + (R + \mathbb{E}^* \bar{R}\mathbb{E}),$$

which is a bounded self-adjoint linear operator from $\mathcal{U}[t, T]$ into itself. Then the above can be further written as

$$J^0(t, 0; u) = \langle \mathcal{M}_t u, u\rangle, \tag{3.3.2}$$

and the cost functional $J(t, \xi; u)$ admits the following representation:

$$J(t, \xi; u) = \langle \mathcal{M}_t u, u \rangle + \langle \mathcal{D}_u J(t, \xi; 0), u \rangle + J(t, \xi; 0),$$
$$\forall (t, \xi) \in \mathscr{D}, \; \forall u \in \mathcal{U}[t, T]. \tag{3.3.3}$$

We see that the cost functional $J(t, \xi; u)$ of the mean-field LQ problem has the same structure as that of Problem (SLQ) presented in Sect. 2 of [48, Chap. 2]. So most results there remain true for Problem (MFLQ). To be more precise, let us first introduce the following relevant assumptions.

(A3) $J^0(t, 0; u) \geqslant 0$ for all $u \in \mathcal{U}[t, T]$.
(A4) There exists a constant $\delta > 0$ such that

$$J^0(t, 0; u) \geqslant \delta \, \mathbb{E} \int_t^T |u(s)|^2 ds, \quad \forall u \in \mathcal{U}[t, T].$$

From (3.3.2) and (3.3.3), we see that (A3) is equivalent to each of the following conditions:

(i) $\langle \mathcal{M}_t u, u \rangle \geqslant 0$ for all $u \in \mathcal{U}[t, T]$.
(ii) The map $u \mapsto J^0(t, \xi; u)$ is convex for every $\xi \in \mathcal{X}_t$.
(iii) The map $u \mapsto J(t, \xi; u)$ is convex for every $\xi \in \mathcal{X}_t$.

Likewise, (A4) is equivalent to each of the following conditions:

(i) There exists a constant $\delta > 0$ such that

$$\langle \mathcal{M}_t u, u \rangle \geqslant \delta \, \mathbb{E} \int_t^T |u(s)|^2 ds, \quad \forall u \in \mathcal{U}[t, T].$$

(ii) The map $u \mapsto J^0(t, \xi; u)$ is uniformly convex for every $\xi \in \mathcal{X}_t$.
(iii) The map $u \mapsto J(t, \xi; u)$ is uniformly convex for every $\xi \in \mathcal{X}_t$.

The following result shows that (A3) is a necessary condition for the finiteness (and open-loop solvability) of Problem (MFLQ) at the initial time t, and that (A4) is a sufficient condition for the open-loop solvability at t.

Proposition 3.3.1 *Let (A1)–(A2) hold and let t be a given initial time. The following statements are true:*

(i) *If Problem (MFLQ) is finite at t, then (A3) must hold.*
(ii) *Suppose that (A4) holds. Then Problem (MFLQ) is uniquely open-loop solvable at t, and the unique optimal control for the initial pair (t, ξ) is given by*

$$u^* = -\frac{1}{2} \mathcal{M}_t^{-1} \mathcal{D}_u J(t, \xi; 0).$$

Moreover,

$$V(t, \xi) = J(t, \xi; 0) - \frac{1}{4} \left| \mathcal{M}_t^{-\frac{1}{2}} \mathcal{D}_u J(t, \xi; 0) \right|^2.$$

Proof The proof is straightforward, and it is omitted.   □

Assume that the necessary condition (A3) for the finiteness of Problem (MFLQ) holds. For $\varepsilon > 0$, let us consider the state Equation (3.1.1) and the following uniformly convex cost functional:

$$
\begin{aligned}
J_\varepsilon(t, \xi; u) &\triangleq J(t, \xi; u) + \varepsilon \mathbb{E} \int_t^T |u(s)|^2 ds \\
&= \langle (\mathcal{M}_t + \varepsilon I)u, u \rangle + \langle \mathcal{D}_u J(t, \xi; 0), u \rangle + J(t, \xi; 0). \quad (3.3.4)
\end{aligned}
$$

Denote the corresponding optimal control problem and value function by Problem (MFLQ)$_\varepsilon$ and $V_\varepsilon(t, \xi)$, respectively. According to Proposition 3.3.1(ii), for any $\xi \in \mathcal{X}_t$, Problem (MFLQ)$_\varepsilon$ admits a unique optimal control

$$
u_\varepsilon^* = -\frac{1}{2}(\mathcal{M}_t + \varepsilon I)^{-1} \mathcal{D}_u J(t, \xi; 0), \quad (3.3.5)
$$

and its value function is given by

$$
V_\varepsilon(t, \xi) = J(t, \xi; 0) - \frac{1}{4} |(\mathcal{M}_t + \varepsilon I)^{-\frac{1}{2}} \mathcal{D}_u J(t, \xi; 0)|^2. \quad (3.3.6)
$$

Theorem 3.3.2 *Let (A1)–(A3) hold, and let $\xi \in \mathcal{X}_t$ be a given initial state. We have the following:*

(i) $\lim_{\varepsilon \to 0} V_\varepsilon(t, \xi) = V(t, \xi)$.

(ii) *The family $\{u_\varepsilon^*\}_{\varepsilon>0}$ defined by (3.3.5) is a minimizing family of the map $u \mapsto J(t, \xi; u)$. That is,*

$$
\lim_{\varepsilon \to 0} J(t, \xi; u_\varepsilon^*) = \inf_{u \in \mathcal{U}[t,T]} J(t, \xi; u) = V(t, \xi). \quad (3.3.7)
$$

(iii) *The following statements are equivalent:*

(a) *Problem (MFLQ) is open-loop solvable at (t, ξ);*

(b) *The family $\{u_\varepsilon^*\}_{\varepsilon>0}$ is bounded in $\mathcal{U}[t, T]$;*

(c) *u_ε^* converges strongly in $\mathcal{U}[t, T]$ as $\varepsilon \to 0$.*

In this case, the strong limit of u_ε^* is an open-loop optimal control of Problem (MFLQ) for (t, ξ).

The proof of Theorem 3.3.2 follows from Proposition 1.3.4 of [48, Chap. 1] (see also Theorem 2.6.2 of [48, Chap. 2]). The details are omitted here.

3.4   Uniform Convexity and Riccati Equations

Theorem 3.3.2 tells us that under the necessary condition (A3), in order to solve
Problem (MFLQ), we need only solve Problem (MFLQ)$_\varepsilon$, which has the uniformly
convex cost functional, then pass to the limit with $\varepsilon \to 0$. When the uniform convexity condition (A4) holds, Proposition 3.3.1 gives a representation of the unique
optimal control u^* in terms of the operator $\mathcal{M}_t$ and $\mathcal{D}_u J(t, \xi; 0)$. However, such a
representation is not convenient for applications, since both $\mathcal{M}_t$ and $\mathcal{D}_u J(t, \xi; 0)$
are in abstract forms and very difficult to compute. To obtain a more explicit form
of the optimal control, we will consider the following two Riccati equations:

$$
\begin{cases}
\dot{P} + \mathcal{Q}(P) - \mathcal{S}(P)^\top \mathcal{R}(P)^\dagger \mathcal{S}(P) = 0, \\
P(T) = G,
\end{cases}
\tag{3.4.1}
$$

$$
\begin{cases}
\dot{\Pi} + \widehat{\mathcal{Q}}(P, \Pi) - \widehat{\mathcal{S}}(P, \Pi)^\top \widehat{\mathcal{R}}(P)^\dagger \widehat{\mathcal{S}}(P, \Pi) = 0, \\
\Pi(T) = \widehat{G},
\end{cases}
\tag{3.4.2}
$$

where in (3.4.1), we have adopted the notation

$$
\begin{aligned}
\mathcal{Q}(P) &= PA + A^\top P + C^\top PC + Q, \quad \mathcal{R}(P) = R + D^\top PD, \\
\mathcal{S}(P) &= B^\top P + D^\top PC + S,
\end{aligned}
\tag{3.4.3}
$$

introduced in (1.1.8), and in (3.4.2), we have employed the following notation:

$$
\begin{aligned}
\widehat{A} &= A + \bar{A}, \quad \widehat{B} = B + \bar{B}, \quad \widehat{C} = C + \bar{C}, \quad \widehat{D} = D + \bar{D}, \\
\widehat{Q} &= Q + \bar{Q}, \quad \widehat{S} = S + \bar{S}, \quad \widehat{R} = R + \bar{R}, \quad \widehat{G} = G + \bar{G}, \\
\widehat{\mathcal{Q}}(P, \Pi) &= \Pi\widehat{A} + \widehat{A}^\top \Pi + \widehat{C}^\top PC + \widehat{Q}, \quad \widehat{\mathcal{R}}(P) = \widehat{R} + \widehat{D}^\top PD, \\
\widehat{\mathcal{S}}(P, \Pi) &= \widehat{B}^\top \Pi + \widehat{D}^\top PC + \widehat{S}.
\end{aligned}
\tag{3.4.4}
$$

It should be noted that in the above the variable s has been suppressed for simplicity.

In this section, we will (i) establish the equivalence between the uniform convexity
of the cost functional and the strongly regular solvability of the above two Riccati
equations, (ii) use the solutions of (3.4.1) and (3.4.2) to construct the optimal control,
and (iii) present some sufficient conditions on the coefficients of the state equation
and the weighting matrices of the cost functional that guarantee the uniform convexity
of the cost functional. To clearly present the results, we will divide the developments
into three subsections.

3.4.1 Solvability of Riccati Equations: Sufficiency of the Uniform Convexity

Our first result of this section is as follows.

Theorem 3.4.1 *Let (A1)–(A2) and (A4) hold. Then the Riccati equation* (3.4.1) *admits a unique solution* $P \in C([t, T]; \mathbb{S}^n)$ *such that*

$$\mathcal{R}(P) \equiv R + D^\top P D \gg 0, \quad \widehat{\mathcal{R}}(P) \equiv \widehat{R} + \widehat{D}^\top P \widehat{D} \gg 0, \tag{3.4.5}$$

and the Riccati equation (3.4.2) *admits a unique solution* $\Pi \in C([t, T]; \mathbb{S}^n)$.

The proof of Theorem 3.4.1 will be divided into two steps. In the first step, we will prove that the Riccati equation (3.4.1) admits a unique solution $P \in C([t, T]; \mathbb{S}^n)$ such that the first condition in (3.4.5) holds. In the second step, we will further show that the solution P satisfies the second condition in (3.4.5), and that the Riccati equation (3.4.2) is solvable.

For the first step, we need the following lemmas.

Lemma 3.4.2 *Let (A1)–(A2) and (A4) hold. Then there exists a constant* $\alpha \in \mathbb{R}$ *such that value function of Problem (MFLQ)^0 satisfies*

$$V^0(s, \xi) \geqslant \alpha \mathbb{E}|\xi|^2, \quad \forall (s, \xi) \in \mathscr{D} \text{ with } s \geqslant t \text{ and } \mathbb{E}\xi = 0.$$

Proof Let $s \in [t, T]$. For a control $u \in \mathcal{U}[s, T]$, we define its *zero-extension* on $[t, T]$ as follows:

$$[01_{[t,s)} \oplus u](r) = \begin{cases} 0, & r \in [t, s), \\ u(r), & r \in [s, T]. \end{cases}$$

It is easily seen that $v \triangleq 01_{[t,s)} \oplus u \in \mathcal{U}[t, T]$. Let $\mathring{X}_0^v$ be the solution to the following mean-field SDE over $[t, T]$:

$$\begin{cases} d\mathring{X}_0^v(r) = \{A(r)\mathring{X}_0^v(r) + \bar{A}(r)\mathbb{E}[\mathring{X}_0^v(r)] + B(r)v(r) + \bar{B}(r)\mathbb{E}[v(r)]\}dr \\ \qquad\qquad + \{C(r)\mathring{X}_0^v(r) + \bar{C}(r)\mathbb{E}[\mathring{X}_0^v(r)] + D(r)v(r) + \bar{D}(r)\mathbb{E}[v(r)]\}dW, \\ \mathring{X}_0^v(t) = 0. \end{cases}$$

Since the initial state is zero, we have $\mathring{X}_0^v(r) = 0$ for $r \in [t, s]$. Hence,

$$J^0(s, 0; u) = J^0(t, 0; 01_{[t,s)} \oplus u)$$

$$\geqslant \delta \mathbb{E} \int_t^T |[01_{[t,s)} \oplus u](r)|^2 dr = \delta \mathbb{E} \int_s^T |u(r)|^2 dr. \tag{3.4.6}$$

Let (X, Y, Z) be the adapted solution to the mean-field FBSDE

$$\begin{cases} dX(r) = \{AX + \bar{A}\mathbb{E}[X]\}dr + \{CX + \bar{C}\mathbb{E}[X]\}dW, \quad r \in [s, T], \\ dY(r) = -\{A^\top Y + \bar{A}^\top \mathbb{E}[Y] + C^\top Z + \bar{C}^\top \mathbb{E}[Z] \\ \qquad\qquad + QX + \bar{Q}\mathbb{E}[X]\}dr + ZdW, \quad r \in [s, T], \\ X(s) = \xi, \quad Y(T) = GX(T) + \bar{G}\mathbb{E}[X(T)], \end{cases}$$

and let $(\mathbb{X}, \mathbb{Y}, \mathbb{Z})$ be the adapted solution to the matrix FBSDE

$$\begin{cases} d\mathbb{X}(r) = A\mathbb{X}dr + C\mathbb{X}dW, \quad r \in [0, T], \\ d\mathbb{Y}(r) = -(A^\top \mathbb{Y} + C^\top \mathbb{Z} + Q\mathbb{X})dr + \mathbb{Z}dW, \quad r \in [0, T], \\ \mathbb{X}(0) = I_n, \quad \mathbb{Y}(T) = G\mathbb{X}(T). \end{cases}$$

According to Proposition 3.2.1, we have

$$J^0(s, \xi; u) - J^0(s, \xi; 0) - J^0(s, 0; u)$$

$$= 2\mathbb{E}\int_s^T \langle B^\top Y + \bar{B}^\top \mathbb{E}[Y] + D^\top Z + \bar{D}^\top \mathbb{E}[Z] + SX + \bar{S}\mathbb{E}[X], u\rangle dr. \quad (3.4.7)$$

If $\mathbb{E}[\xi] = 0$, then

$$\mathbb{E}[X(r)] = 0, \quad \forall r \in [s, T], \tag{3.4.8}$$

and one can easily verify that for $r \in [s, T]$,

$$(X(r), Y(r), Z(r)) = (\mathbb{X}(r)\mathbb{X}(s)^{-1}, \mathbb{Y}(r)\mathbb{X}(s)^{-1}, \mathbb{Z}(r)\mathbb{X}(s)^{-1})\xi. \tag{3.4.9}$$

Since for $r \in [s, T]$, $\mathbb{X}(r)\mathbb{X}(s)^{-1}$, $\mathbb{Y}(r)\mathbb{X}(s)^{-1}$ and $\mathbb{Z}(r)\mathbb{X}(s)^{-1}$ are independent of $\mathcal{F}_s$, we have

$$\mathbb{E}[X(r)] = \mathbb{E}[Y(r)] = \mathbb{E}[Z(r)] = 0, \quad \forall r \in [s, T],$$

and thereby (3.4.7) reduces to

$$J^0(s, \xi; u) - J^0(s, \xi; 0) - J^0(s, 0; u) = 2\mathbb{E}\int_s^T \langle B^\top Y + D^\top Z + SX, u\rangle dr.$$

Now using the Cauchy-Schwarz inequality and (3.4.6), we obtain

$$J^0(s, \xi; u) \geqslant J^0(s, \xi; 0) + J^0(s, 0; u) - \delta\mathbb{E}\int_s^T |u(r)|^2 dr$$

$$- \frac{1}{\delta}\mathbb{E}\int_s^T |B^\top Y + D^\top Z + SX|^2 dr$$

$$\geqslant J^0(s, \xi; 0) - \frac{1}{\delta}\mathbb{E}\int_s^T |B^\top Y + D^\top Z + SX|^2 dr. \tag{3.4.10}$$

Recalling (3.4.8) and (3.4.9), we can rewrite $J^0(s, \xi; 0)$ as

$$J^0(s, \xi; 0) = \mathbb{E}\left\{ \langle GX(T), X(T) \rangle + \int_s^T \langle Q(r)X(r), X(r) \rangle dr \right\}$$
$$= \mathbb{E}\left\{ \xi^\top [\mathbb{X}(s)^{-1}]^\top \mathbb{X}(T)^\top G\mathbb{X}(T)\mathbb{X}(s)^{-1}\xi \right.$$
$$\left. + \xi^\top \left(\int_s^T [\mathbb{X}(s)^{-1}]^\top \mathbb{X}(r)^\top Q(r)\mathbb{X}(r)\mathbb{X}(s)^{-1} dr \right) \xi \right\}.$$

Similarly,

$$\mathbb{E} \int_s^T |B(r)^\top Y(r) + D(r)^\top Z(r) + S(r)X(r)|^2 dr$$
$$= \mathbb{E}\left\{ \xi^\top \left(\int_s^T [\mathbb{X}(s)^{-1}]^\top [B(r)^\top Y(r) + D(r)^\top Z(r) + S(r)X(r)]^\top \right. \right.$$
$$\left. \left. \times [B(r)^\top Y(r) + D(r)^\top Z(r) + S(r)X(r)]\mathbb{X}(s)^{-1} dr \right) \xi \right\}.$$

Thus, with the notation

$$\mathbb{M}(s) = \mathbb{E}\left\{ [\mathbb{X}(s)^{-1}]^\top \mathbb{X}(T)^\top G\mathbb{X}(T)\mathbb{X}(s)^{-1} \right.$$
$$+ \int_s^T [\mathbb{X}(s)^{-1}]^\top \mathbb{X}(r)^\top Q(r)\mathbb{X}(r)\mathbb{X}(s)^{-1} dr$$
$$- \frac{1}{\delta} \int_s^T [\mathbb{X}(s)^{-1}]^\top [B(r)^\top Y(r) + D(r)^\top Z(r) + S(r)X(r)]^\top$$
$$\left. \times [B(r)^\top Y(r) + D(r)^\top Z(r) + S(r)X(r)]\mathbb{X}(s)^{-1} dr \right\},$$

(3.4.10) becomes
$$J^0(s, \xi; u) \geqslant \mathbb{E}[\xi^\top \mathbb{M}(s)\xi].$$

The desired result then follows from the fact that the function $\mathbb{M} : [t, T] \to \mathbb{S}^n$ is continuous. $\qquad \square$

Lemma 3.4.3 *Let (A1)–(A2) hold. For $\Theta \in \boldsymbol{\Theta}[t, T]$, let $P_\Theta \in C([t, T]; \mathbb{S}^n)$ denote the solution to the following Lyapunov equation:*

$$\begin{cases} \dot{P}_\Theta + P_\Theta(A + B\Theta) + (A + B\Theta)^\top P_\Theta + (C + D\Theta)^\top P_\Theta(C + D\Theta) \\ \qquad + \Theta^\top R\Theta + S^\top \Theta + \Theta^\top S + Q = 0, \\ P_\Theta(T) = G. \end{cases} \qquad (3.4.11)$$

If there exist constants $\alpha \in \mathbb{R}$ and $\beta > 0$ such that for all $\Theta \in \boldsymbol{\Theta}[t, T]$,

$$P_{\Theta}(s) \geqslant \alpha I_n, \quad R(s) + D(s)^{\top} P_{\Theta}(s) D(s) \geqslant \beta I_m, a.e. \ s \in [t, T], \qquad (3.4.12)$$

then the Riccati equation (3.4.1) is strongly regularly solvable.

Proof The proof is the same as the proof of Chap. 2, Theorem 2.5.6, part (i) $\Rightarrow$ (ii) in [48]. $\qquad\qquad\qquad\qquad\qquad\qquad\qquad\qquad\qquad\qquad\qquad\qquad\qquad\qquad\qquad\qquad\square$

Proof of Theorem 3.4.1. *Step 1:* We need only show that the condition stated in Lemma 3.4.3 holds. To this end, let $\Theta \in \boldsymbol{\Theta}[t, T]$ and denote simply by P the corresponding solution of (3.4.11). Take an arbitrary deterministic function $u \in L^2(t, T; \mathbb{R}^m) \subseteq \mathcal{U}[t, T]$, and let X^u denote the solution to the following SDE over $[t, T]$:

$$\begin{cases} dX^u(s) = [(A + B\Theta)X^u + BuW]ds + [(C + D\Theta)X^u + DuW]dW, \\ X^u(t) = 0. \end{cases}$$

One sees that

$$v \triangleq \Theta X^u + uW \in \mathcal{U}[t, T], \qquad\qquad\qquad (3.4.13)$$

and that

$$\mathbb{E}[X^u(s)] = 0, \quad \mathbb{E}[v(s)] = 0, \quad \forall s \in [t, T]. \qquad\qquad (3.4.14)$$

Thus, X^u also satisfies the mean-field SDE

$$\begin{cases} dX^u(s) = \{AX^u + \bar{A}\mathbb{E}[X^u] + Bv + \bar{B}\mathbb{E}[v]\}ds \\ \qquad\qquad + \{CX^u + \bar{C}\mathbb{E}[X^u] + Dv + \bar{D}\mathbb{E}[v]\}dW, \quad s \in [t, T], \\ X^u(t) = 0. \end{cases}$$

This means that

$$J^0(t, 0; v) = \mathbb{E}\langle GX^u(T), X^u(T)\rangle$$
$$+ \mathbb{E}\int_t^T \Big[\langle QX^u, X^u\rangle + 2\langle SX^u, v\rangle + \langle Rv, v\rangle\Big]ds. \qquad (3.4.15)$$

By applying Itô's formula to $s \to \langle P(s)X^u(s), X^u(s)\rangle$, we obtain

$$\mathbb{E}\langle GX^u(T), X^u(T)\rangle$$

$$= \mathbb{E}\int_t^T \left\{ \langle \dot{P}X^u, X^u\rangle + \langle P[(A+B\Theta)X^u + BuW], X^u\rangle \right.$$

$$+ \langle PX^u, (A+B\Theta)X^u + BuW\rangle$$

$$\left. + \langle P[(C+D\Theta)X^u + DuW], (C+D\Theta)X^u + DuW\rangle \right\}ds$$

$$= \mathbb{E}\int_t^T \left\{ -\langle (\Theta^\top R\Theta + S^\top\Theta + \Theta^\top S + Q)X^u, X^u\rangle \right.$$

$$\left. + 2\langle [B^\top P + D^\top P(C+D\Theta)]X^u, uW\rangle + \langle D^\top PDuW, uW\rangle \right\}ds.$$

Substituting (3.4.13) and the above into (3.4.15) gives

$$J^0(t, 0; v) = \mathbb{E}\int_t^T \left\{ 2\langle[\mathcal{S}(P)+\mathcal{R}(P)\Theta]X^u, uW\rangle + \langle \mathcal{R}(P)uW, uW\rangle \right\}ds.$$
$$(3.4.16)$$

By the assumption (A4),

$$J^0(t, 0; v) \geqslant \delta\mathbb{E}\int_t^T |v(s)|^2 ds = \delta\mathbb{E}\int_t^T |\Theta(s)X^u(s) + u(s)W(s)|^2 ds. \quad (3.4.17)$$

Combining (3.4.16) and (3.4.17) yields

$$\mathbb{E}\int_t^T \left\{ 2\langle[\mathcal{S}(P)+(\mathcal{R}(P)-\delta I_m)\Theta]WX^u, u\rangle + W^2\langle[\mathcal{R}(P)-\delta I_m]u, u\rangle \right\}ds$$

$$= \delta\mathbb{E}\int_t^T |\Theta(s)X^u(s)|^2 ds \geqslant 0. \quad (3.4.18)$$

For simplicity of notation, let us write

$$\Delta = \mathcal{S}(P) + [\mathcal{R}(P) - \delta I_m]\Theta, \quad \Lambda = \mathcal{R}(P) - \delta I_m.$$

Noting that u is a deterministic function, we can rewrite (3.4.18) as

$$\int_t^T \left\{ 2\langle \Delta(s)\mathbb{E}[W(s)X^u(s)], u(s)\rangle + s\langle \Lambda(s)u(s), u(s)\rangle \right\}ds \geqslant 0. \quad (3.4.19)$$

By applying Itô's formula to $s \mapsto W(s)X^u(s)$ and taking expectations, we see that $V(s) \triangleq \mathbb{E}[W(s)X^u(s)]$ satisfies the following ODE:

$$\begin{cases} \dot{V}(s) = [A(s)+B(s)\Theta(s)]V(s) + sB(s)u(s), & s\in[t, T], \\ V(t) = 0. \end{cases}$$

Now we take u to be the form

$$u(s) = u_0 1_{[t', t'+h]}(s),$$

where $u_0 \in \mathbb{R}^m$ is a constant vector and $t \leqslant t' < t' + h \leqslant T$. Then

$$\mathbb{E}[W(s)X^u(s)] = \begin{cases} 0, & s \in [t, t'], \\ \Phi(s) \int_{t'}^{s \wedge (t'+h)} \Phi(r)^{-1} B(r) r u_0 dr, & s \in [t', T], \end{cases}$$

where Φ is the fundamental matrix for the system

$$\dot{x}(s) = [A(s) + B(s)\Theta(s)]x(s), \quad s \in [0, T].$$

Consequently, (3.4.19) becomes

$$\int_{t'}^{t'+h} \left[2 \langle \Delta(s)\Phi(s) \int_{t'}^{s} \Phi(r)^{-1} B(r) r u_0 dr, u_0 \rangle + s \langle \Lambda(s) u_0, u_0 \rangle \right] ds \geqslant 0.$$

Dividing both sides by h and letting $h \to 0$, we obtain

$$t' \langle \Lambda(t') u_0, u_0 \rangle \geqslant 0, \quad \text{a.e. } t' \in [t, T].$$

Since $u_0 \in \mathbb{R}^m$ is arbitrary, the above implies that $\Lambda(s) \geqslant 0$ for almost every $s \in [t, T]$, or equivalently,

$$R(s) + D(s)^\top P(s) D(s) \geqslant \delta I_m, \quad \text{a.e. } s \in [t, T].$$

To show that

$$P(s) \geqslant \alpha I_n, \quad \text{a.e. } s \in [t, T]$$

for some constant α independent of Θ, we denote by X the solution of

$$\begin{cases} dX(r) = (A + B\Theta)X dr + (C + D\Theta)X dW, & r \in [s, T], \\ X(s) = W(s)x, \end{cases}$$

and set $w = \Theta X \in \mathcal{U}[s, T]$. Similar to the previous arguments, by applying Itô's formula to $r \to \langle P(r)X(r), X(r) \rangle$, we can derive that

$$J^0(s, W(s)x; w) = \mathbb{E}\langle P(s)W(s)x, W(s)x \rangle = s \langle P(s)x, x \rangle.$$

Then with α being the constant in Lemma 3.4.2,

$$s\langle P(s)x, x\rangle \geqslant V^0(s, W(s)x) \geqslant \alpha\mathbb{E}|W(s)x|^2 = s\alpha|x|^2, \quad \forall (s, x) \in [t, T] \times \mathbb{R}^n,$$

which further implies that $P(s) \geqslant \alpha I_n$ for all $s \in [t, T]$. $\qquad\square$

For the second step of the proof of Theorem 3.4.1, we introduce a deterministic LQ optimal control problem. With the notation of (3.4.4), let us consider the state equation

$$\begin{cases} \dot{y}(s) = \widehat{A}(s)y(s) + \widehat{B}(s)u(s), & s \in [t, T], \\ y(t) = x, \end{cases} \tag{3.4.20}$$

and the cost functional

$$\widehat{J}(t, x; u) = \langle \widehat{G}y(T), y(T)\rangle + \int_t^T \left\langle \begin{pmatrix} \widehat{Q}(P, 0) & \widehat{S}(P, 0)^\top \\ \widehat{S}(P, 0) & \widehat{R}(P) \end{pmatrix} \begin{pmatrix} y \\ u \end{pmatrix}, \begin{pmatrix} y \\ u \end{pmatrix} \right\rangle ds.$$

We pose the following problem.

Problem (DLQ). For given initial pair $(t, x) \in [0, T) \times \mathbb{R}^n$, find a control $u^* \in L^2(t, T; \mathbb{R}^m)$ such that

$$\widehat{J}(t, x; u^*) = \inf_{u \in L^2(t,T;\mathbb{R}^m)} \widehat{J}(t, x; u).$$

Note that the Riccati equation associated with this Problem (DLQ) is exactly (3.4.2). Thus, the second step will be accomplished once we prove the following result.

Theorem 3.4.4 *Let (A1)–(A2) and (A4) hold. Then the map $u \mapsto \widehat{J}(t, 0; u)$ is uniformly convex, i.e., there exists a constant $\lambda > 0$ such that*

$$\widehat{J}(t, 0; u) \geqslant \lambda \int_t^T |u(s)|^2 ds, \quad \forall u \in L^2(t, T; \mathbb{R}^m).$$

Hence, the strongly regular solution P of the Riccati equation (3.4.1) satisfies

$$\widehat{R}(P) \equiv \widehat{R} + \widehat{D}^\top P \widehat{D} \gg 0,$$

and the Riccati equation (3.4.2) admits a unique solution $\Pi \in C([t, T]; \mathbb{S}^n)$.

Proof Let $\Theta = -\mathcal{R}(P)^{-1}\mathcal{S}(P)$. We claim that

$$J^0(t, 0; \Theta X + v) = \widehat{J}(t, 0; \Theta y + v), \quad \forall v \in L^2(t, T; \mathbb{R}^m). \tag{3.4.21}$$

To prove (3.4.21), fix an arbitrary $v \in L^2(t, T; \mathbb{R}^m)$. Let y be the solution of

$$\begin{cases} \dot{y}(s) = \widehat{A}(s)y(s) + \widehat{B}(s)[\Theta(s)y(s) + v(s)], & s \in [t, T], \\ y(t) = 0, \end{cases}$$

and X be the solution of

$$\begin{cases} dX(s) = \big\{ AX + \bar{A}\mathbb{E}[X] + B(\Theta X + v) + \bar{B}\mathbb{E}[\Theta X + v] \big\} ds \\ \qquad\quad + \big\{ CX + \bar{C}\mathbb{E}[X] + D(\Theta X + v) + \bar{D}\mathbb{E}[\Theta X + v] \big\} dW, & s \in [t, T], \\ X(t) = 0. \end{cases}$$

Since Θ and v are deterministic, we have

$$\begin{cases} d\mathbb{E}[X(s)] = \big\{ \widehat{A}\mathbb{E}[X] + \widehat{B}(\Theta\mathbb{E}[X] + v) \big\} ds, & s \in [t, T], \\ \mathbb{E}[X(t)] = 0. \end{cases}$$

Thus, the functions $s \mapsto \mathbb{E}[X(s)]$ and $s \mapsto y(s)$ satisfy the same ODE, and hence by the uniqueness of solutions,

$$\mathbb{E}[X(s)] = y(s), \quad s \in [t, T].$$

Let

$$z(s) \triangleq X(s) - \mathbb{E}[X(s)] = X(s) - y(s).$$

Then

$$\mathbb{E}\langle GX(T), X(T)\rangle + \langle \bar{G}\mathbb{E}[X(T)], \mathbb{E}[X(T)]\rangle = \mathbb{E}\langle Gz(T)z(T)\rangle + \langle \widehat{G}y(T), y(T)\rangle,$$
$$\mathbb{E}\langle QX, X\rangle + \langle \bar{Q}\mathbb{E}[X], \mathbb{E}[X]\rangle = \mathbb{E}\langle Qz, z\rangle + \langle \widehat{Q}y, y\rangle.$$

Also, let

$$u(s) \triangleq \Theta(s)X(s) + v(s).$$

Then

$$\mathbb{E}[u(s)] = \Theta(s)y(s) + v(s).$$

Consequently,

$$\mathbb{E}\langle SX, u\rangle + \langle \bar{S}\mathbb{E}[X], \mathbb{E}[u]\rangle = \mathbb{E}\langle Sz, \Theta z\rangle + \langle \widehat{S}y, \Theta y + v\rangle,$$
$$\mathbb{E}\langle Ru, u\rangle + \langle \bar{R}\mathbb{E}[u], \mathbb{E}[u]\rangle = \mathbb{E}\langle R\Theta z, \Theta z\rangle + \langle \widehat{R}(\Theta y + v), \Theta y + v\rangle.$$

It follows that

$$J^0(t, 0; \Theta X + v)$$

$$= \mathbb{E}\left\{ \langle Gz(T)z(T) \rangle + \int_t^T \left[\langle Qz, z \rangle + 2\langle Sz, \Theta z \rangle + \langle R\Theta z, \Theta z \rangle \right] ds \right\}$$

$$+ \langle \widehat{G}y(T), y(T) \rangle + \int_t^T \left[\langle \widehat{Q}y, y \rangle + 2\langle \widehat{S}y, \Theta y + v \rangle \right.$$

$$+ \langle \widehat{R}(\Theta y + v), \Theta y + v \rangle \Big] ds. \tag{3.4.22}$$

Observe that z satisfies the following SDE:

$$\begin{cases} dz(s) = (A + B\Theta)z \, ds + \left[(C + D\Theta)z + \widehat{C}y + \widehat{D}(\Theta y + v) \right] dW, \\ z(t) = 0. \end{cases}$$

Also, keep in mind that v is deterministic and $\mathbb{E}[z(s)] \equiv 0$, and note that

$$0 = \dot{P} + P(A + B\Theta) + (A + B\Theta)^\top P + (C + D\Theta)^\top P(C + D\Theta)$$

$$+ \Theta^\top R\Theta + S^\top \Theta + \Theta^\top S + Q.$$

Then by applying Itô's formula to $s \mapsto \langle P(s)z(s), z(s) \rangle$, we can obtain

$$\mathbb{E}\left\{ \langle Gz(T)z(T) \rangle + \int_t^T \left[\langle Qz, z \rangle + 2\langle Sz, \Theta z \rangle + \langle R\Theta z, \Theta z \rangle \right] ds \right\}$$

$$= \int_t^T \left[\langle \widehat{C}^\top P\widehat{C}y, y \rangle + 2\langle \widehat{D}^\top P\widehat{C}y, \Theta y + v \rangle \right.$$

$$+ \langle \widehat{D}^\top P\widehat{D}(\Theta y + v), \Theta y + v \rangle \Big] ds. \tag{3.4.23}$$

Substituting (3.4.23) into (3.4.22) gives (3.4.21). Consequently, we have by the assumption (A4) that

$$\widehat{J}(t, 0; \Theta y + v) \geqslant \delta \mathbb{E} \int_t^T |\Theta(s)X(s) + v(s)|^2 ds$$

$$\geqslant \delta \int_t^T |\mathbb{E}[\Theta(s)X(s) + v(s)]|^2 ds = \delta \int_t^T |\Theta(s)y(s) + v(s)|^2 ds.$$

Since $v \in L^2(t, T; \mathbb{R}^m)$ is arbitrary, the uniform convexity of $u \mapsto \widehat{J}(t, 0; u)$ follows. The rest assertions are immediate consequences of Theorem 1.1.15 (with the initial time 0 replaced by t). $\qquad\square$

3.4.2 Solvability of Riccati Equations: Necessity of the Uniform Convexity

We have shown in Theorem 3.4.1 that the uniform convexity condition (A4) implies the solvability of the Riccati equations (3.4.1) and (3.4.2). In this section we establish the converse to Theorem 3.4.1.

First we need the following lemma.

Lemma 3.4.5 *Let (A1)–(A2) hold. For $u \in \mathcal{U}[t, T]$, let X_0^u denote the solution to*

$$
\begin{cases}
dX_0^u(s) = \big\{ AX_0^u + \bar{A}\mathbb{E}[X_0^u] + Bu + \bar{B}\mathbb{E}[u] \big\}ds \\
\qquad\qquad + \big\{ CX_0^u + \bar{C}\mathbb{E}[X_0^u] + Du + \bar{D}\mathbb{E}[u] \big\}dW, \quad s \in [t, T], \quad (3.4.24) \\
X_0^u(t) = 0.
\end{cases}
$$

Then for every $\Theta, \bar{\Theta} \in \boldsymbol{\Theta}[t, T]$, there exists a constant $\gamma > 0$ such that for all $u \in \mathcal{U}[t, T]$,

$$
\mathbb{E}\int_t^T |u(s) - \Theta(s)\{X_0^u(s) - \mathbb{E}[X_0^u(s)]\}|^2 ds \geqslant \gamma \mathbb{E}\int_t^T |u(s)|^2 ds, \qquad (3.4.25)
$$

$$
\int_t^T |\mathbb{E}[u(s)] - \bar{\Theta}(s)\mathbb{E}[X_0^u(s)]|^2 ds \geqslant \gamma \int_t^T |\mathbb{E}[u(s)]|^2 ds. \qquad (3.4.26)
$$

Proof For each $\Theta \in \boldsymbol{\Theta}[t, T]$, we can define a bounded linear operator $\mathcal{A} : \mathcal{U}[t, T] \to \mathcal{U}[t, T]$ by

$$
\mathcal{A}u = u - \Theta(X_0^u - \mathbb{E}[X_0^u]).
$$

The operator $\mathcal{A}$ is bijective and its inverse $\mathcal{A}^{-1}$ is given by

$$
\mathcal{A}^{-1}v = v + \Theta\big(\widetilde{X}_0^v - \mathbb{E}[\widetilde{X}_0^v]\big),
$$

where $\widetilde{X}_0^v = \{\widetilde{X}_0^v(s); t \leqslant s \leqslant T\}$ is the solution of

$$
\begin{cases}
d\widetilde{X}_0^v(s) = \big\{ (A + B\Theta)\widetilde{X}_0^v + (\bar{A} - B\Theta)\mathbb{E}[\widetilde{X}_0^v] + Bv + \bar{B}\mathbb{E}[v] \big\}ds \\
\qquad\qquad + \big\{ (C + D\Theta)\widetilde{X}_0^v + (\bar{C} - D\Theta)\mathbb{E}[\widetilde{X}_0^v] + Dv + \bar{D}\mathbb{E}[v] \big\}dW, \\
\widetilde{X}_0^v(t) = 0.
\end{cases}
$$

By the open mapping theorem, $\mathcal{A}^{-1}$ is also a bounded operator with norm $\|\mathcal{A}^{-1}\| > 0$. Thus, for each $u \in \mathcal{U}[t, T]$,

$$\mathbb{E}\int_t^T |u(s)|^2 ds = \mathbb{E}\int_t^T |(\mathcal{A}^{-1}\mathcal{A}u)(s)|^2 ds \leqslant \|\mathcal{A}^{-1}\| \mathbb{E}\int_t^T |(\mathcal{A}u)(s)|^2 ds$$

$$= \|\mathcal{A}^{-1}\| \mathbb{E}\int_t^T |u(s) - \Theta(s)\{X_0^u(s) - \mathbb{E}[X_0^u(s)]\}|^2 ds.$$

This shows that (3.4.25) holds with $\gamma = \|\mathcal{A}^{-1}\|^{-1}$.

To prove (3.4.26), for each deterministic $v \in L^2(t, T; \mathbb{R}^{m\times n})$, let y^v denote the solution to

$$\begin{cases} \dot{y}(s) = \widehat{A}(s)y(s) + \widehat{B}(s)v(s), & s \in [t, T], \\ y(t) = 0. \end{cases} \tag{3.4.27}$$

For every $\bar{\Theta} \in \boldsymbol{\Theta}[t, T]$, we can define a bounded linear operator $\mathcal{B}: L^2(t, T; \mathbb{R}^m) \to L^2(t, T; \mathbb{R}^m)$ by

$$\mathcal{B}v = v - \bar{\Theta}y^v.$$

Similar to the previous argument, we can show that $\mathcal{B}$ is invertible and that

$$\int_t^T |v(s) - \bar{\Theta}(s)y^v(s)|^2 ds \geqslant \frac{1}{\|\mathcal{B}^{-1}\|} \int_t^T |v(s)|^2 ds, \quad \forall v \in L^2(t, T; \mathbb{R}^{m\times n}).$$

Since $\mathbb{E}[X_0^u]$ satisfies (3.4.27) with $v = \mathbb{E}[u]$, (3.4.26) follows. $\qquad\square$

Now we state and prove the converse to Theorem 3.4.1.

Theorem 3.4.6 *Let (A1)–(A2) hold. Suppose that the Riccati equation (3.4.1) admits a solution $P \in C([t, T]; \mathbb{S}^n)$ such that (3.4.5) holds and that the Riccati equation (3.4.2) admits a solution $\Pi \in C([t, T]; \mathbb{S}^n)$. Then the uniform convexity condition (A4) holds.*

Proof Recall the notation

$$\begin{aligned} \mathcal{R}(P) &= R + D^\top P D, & \mathcal{S}(P) &= B^\top P + D^\top P C + S, \\ \widehat{\mathcal{R}}(P) &= \widehat{R} + \widehat{D}^\top P \widehat{D}, & \widehat{\mathcal{S}}(P, \Pi) &= \widehat{B}^\top \Pi + \widehat{D}^\top P \widehat{C} + \widehat{S}, \end{aligned}$$

and set

$$\Theta \triangleq -\mathcal{R}^{-1}(P)\mathcal{S}(P), \quad \widehat{\Theta} \triangleq -\widehat{\mathcal{R}}(P)^{-1}\widehat{\mathcal{S}}(P, \Pi). \tag{3.4.28}$$

Let (η, ζ) and $\bar{\eta}$ be the solutions to the BSDE

$$\begin{cases} d\eta(s) = -\big[(A + B\Theta)^\top \eta + (C + D\Theta)^\top \zeta + (C + D\Theta)^\top P\sigma \\ \qquad\qquad + \Theta^\top \rho + Pb + q\big]ds + \zeta dW, & s \in [t, T], \\ \eta(T) = g, \end{cases}$$

and the ODE

$$
\begin{cases}
\dot{\bar{\eta}} + (\widehat{A} + \widehat{B}\Theta)^{\top}\bar{\eta} + \Theta^{\top}\{\widehat{D}^{\top}(P\mathbb{E}[\sigma] + \mathbb{E}[\zeta]) + \mathbb{E}[\rho] + \bar{\rho}\} \\
\quad + \widehat{C}^{\top}(P\mathbb{E}[\sigma] + \mathbb{E}[\zeta]) + \mathbb{E}[q] + \bar{q} + \Pi\mathbb{E}[b] = 0, \quad s \in [t, T], \\
\bar{\eta}(T) = \mathbb{E}[g] + \bar{g},
\end{cases}
$$

respectively, and set

$$
\begin{aligned}
\varphi &= -\mathcal{R}(P)^{-1}\{B^{\top}\eta + D^{\top}(\zeta + P\sigma) + \rho\}, \\
\bar{\varphi} &= -\widehat{\mathcal{R}}(P)^{-1}\{\widehat{B}^{\top}\bar{\eta} + \widehat{D}^{\top}(\mathbb{E}[\zeta] + P\mathbb{E}[\sigma]) + \mathbb{E}[\rho] + \bar{\rho}\}.
\end{aligned}
\tag{3.4.29}
$$

For any $\xi \in L^2_{\mathcal{F}_t}(\Omega; \mathbb{R}^n)$ and $u \in \mathcal{U}[t, T]$, let $X = \{X(s); t \leqslant s \leqslant T\}$ be the corresponding solution to the mean-field state equation (3.1.1), and let

$$
z(s) = X(s) - \mathbb{E}[X(s)], \quad v(s) = u(s) - \mathbb{E}[u(s)], \quad y(s) = \mathbb{E}[X(s)]; \quad t \leqslant s \leqslant T.
$$

Then z satisfies

$$
\begin{cases}
dz(s) = \{Az + Bv + b - \mathbb{E}[b]\}ds \\
\qquad\quad + \{Cz + Dv + \sigma + \widehat{C}y + \widehat{D}\mathbb{E}[u]\}dW, \quad s \in [t, T], \\
z(t) = \xi - \mathbb{E}[\xi],
\end{cases}
$$

and y satisfies

$$
\dot{y}(s) = \widehat{A}y + \widehat{B}\mathbb{E}[u] + \mathbb{E}[b], \quad s \in [t, T]; \qquad y(t) = \mathbb{E}[\xi].
$$

Rewrite the cost functional as follows:

$$
\begin{aligned}
J(t, \xi; u) = \mathbb{E}\Bigg\{ &\langle Gz(T) + 2g, z(T)\rangle \\
&+ \int_t^T \left[\left\langle \begin{pmatrix} Q & S^{\top} \\ S & R \end{pmatrix}\begin{pmatrix} z \\ v \end{pmatrix}, \begin{pmatrix} z \\ v \end{pmatrix}\right\rangle + 2\left\langle \begin{pmatrix} q \\ \rho \end{pmatrix}, \begin{pmatrix} z \\ v \end{pmatrix}\right\rangle \right] ds \Bigg\} \\
&+ \langle \widehat{G}y(T) + 2(\mathbb{E}[g] + \bar{g}), y(T)\rangle \\
&+ \int_t^T \left[\left\langle \begin{pmatrix} \widehat{Q} & \widehat{S}^{\top} \\ \widehat{S} & \widehat{R} \end{pmatrix}\begin{pmatrix} y \\ \mathbb{E}[u] \end{pmatrix}, \begin{pmatrix} y \\ \mathbb{E}[u] \end{pmatrix}\right\rangle \right. \\
&\qquad\qquad \left. + 2\left\langle \begin{pmatrix} \mathbb{E}[q] + \bar{q} \\ \mathbb{E}[\rho] + \bar{\rho} \end{pmatrix}, \begin{pmatrix} y \\ \mathbb{E}[u] \end{pmatrix}\right\rangle \right] ds.
\end{aligned}
\tag{3.4.30}
$$

By applying Itô's formula to $s \mapsto \langle P(s)z(s) + 2\eta(s), z(s)\rangle$ and noting that

$$\mathbb{E}[z(s)] = 0, \quad \mathbb{E}[v(s)] = 0; \quad t \leqslant s \leqslant T,$$

we obtain

$$\mathbb{E}\langle Gz(T) + 2g, z(T)\rangle - \mathbb{E}\langle P(t)(\xi - \mathbb{E}[\xi]) + 2\eta(t), \xi - \mathbb{E}[\xi]\rangle$$
$$= \mathbb{E}\int_t^T \Big[\langle(\dot{P} + PA + A^\top P + C^\top PC)z, z\rangle + 2\langle(PB + C^\top PD)v, z\rangle$$
$$+ \langle D^\top PDv, v\rangle + 2\langle B^\top \eta + D^\top \zeta + D^\top P\sigma, v - \Theta z\rangle - 2\langle\Theta^\top \rho + q, z\rangle$$
$$+ 2\langle P\mathbb{E}[\sigma] + \mathbb{E}[\zeta], \widehat{C}y + \widehat{D}\mathbb{E}[u]\rangle + \langle P(\widehat{C}y + \widehat{D}\mathbb{E}[u]), \widehat{C}y + \widehat{D}\mathbb{E}[u]\rangle$$
$$+ \langle P\sigma, \sigma\rangle + 2\langle\eta, b - \mathbb{E}[b]\rangle + 2\langle\zeta, \sigma\rangle\Big]ds.$$

It follows that

$$\mathbb{E}\Big\{\langle Gz(T) + 2g, z(T)\rangle + \int_t^T \Big[\Big\langle\begin{pmatrix} Q & S^\top \\ S & R \end{pmatrix}\begin{pmatrix} z \\ v \end{pmatrix}, \begin{pmatrix} z \\ v \end{pmatrix}\Big\rangle + 2\Big\langle\begin{pmatrix} q \\ \rho \end{pmatrix}, \begin{pmatrix} z \\ v \end{pmatrix}\Big\rangle\Big]ds\Big\}$$

$$= \mathbb{E}\langle P(t)(\xi - \mathbb{E}[\xi]) + 2\eta(t), \xi - \mathbb{E}[\xi]\rangle + \mathbb{E}\int_t^T \Big[\langle\Theta^\top\mathcal{R}(P)\Theta z, z\rangle$$
$$- 2\langle\Theta^\top\mathcal{R}(P)v, z\rangle + \langle\mathcal{R}(P)v, v\rangle - 2\langle\mathcal{R}(P)\varphi, v - \Theta z\rangle$$
$$+ \langle\widehat{C}^\top P\widehat{C}y, y\rangle + 2\langle\widehat{C}^\top P\widehat{D}\mathbb{E}[u], y\rangle + \langle\widehat{D}^\top P\widehat{D}\mathbb{E}[u], \mathbb{E}[u]\rangle$$
$$+ 2\langle\widehat{C}^\top\big(P\mathbb{E}[\sigma] + \mathbb{E}[\zeta]\big), y\rangle + 2\langle\widehat{D}^\top\big(P\mathbb{E}[\sigma] + \mathbb{E}[\zeta]\big), \mathbb{E}[u]\rangle$$
$$+ \langle P\sigma, \sigma\rangle + 2\langle\eta, b - \mathbb{E}[b]\rangle + 2\langle\zeta, \sigma\rangle\Big]ds$$

$$= \mathbb{E}\langle P(t)(\xi - \mathbb{E}[\xi]) + 2\eta(t), \xi - \mathbb{E}[\xi]\rangle$$
$$+ \mathbb{E}\int_t^T \Big[\langle\mathcal{R}(P)(v - \Theta z - \varphi), v - \Theta z - \varphi\rangle - \langle\mathcal{R}(P)\varphi, \varphi\rangle$$
$$+ \langle\widehat{C}^\top P\widehat{C}y, y\rangle + 2\langle\widehat{C}^\top P\widehat{D}\mathbb{E}[u], y\rangle + \langle\widehat{D}^\top P\widehat{D}\mathbb{E}[u], \mathbb{E}[u]\rangle$$
$$+ 2\langle\widehat{C}^\top\big(P\mathbb{E}[\sigma] + \mathbb{E}[\zeta]\big), y\rangle + 2\langle\widehat{D}^\top\big(P\mathbb{E}[\sigma] + \mathbb{E}[\zeta]\big), \mathbb{E}[u]\rangle$$
$$+ \langle P\sigma, \sigma\rangle + 2\langle\eta, b - \mathbb{E}[b]\rangle + 2\langle\zeta, \sigma\rangle\Big]ds. \tag{3.4.31}$$

Similarly, by applying the integration by parts formula to $s \mapsto \langle\Pi(s)y(s) + 2\bar{\eta}(s), y(s)\rangle$, we can obtain

$$\langle \widehat{G} y(T) + 2(\mathbb{E}[g] + \bar{g}), y(T) \rangle + \int_t^T \left[\left\langle \begin{pmatrix} \widehat{Q} & \widehat{S}^\top \\ \widehat{S} & \widehat{R} \end{pmatrix} \begin{pmatrix} y \\ \mathbb{E}[u] \end{pmatrix}, \begin{pmatrix} y \\ \mathbb{E}[u] \end{pmatrix} \right\rangle \right.$$

$$\left. + 2 \left\langle \begin{pmatrix} \mathbb{E}[q] + \bar{q} \\ \mathbb{E}[\rho] + \bar{\rho} \end{pmatrix}, \begin{pmatrix} y \\ \mathbb{E}[u] \end{pmatrix} \right\rangle \right] ds$$

$$= \langle \Pi(t) \mathbb{E}[\xi] + 2\bar{\eta}(t), \mathbb{E}[\xi] \rangle + \int_t^T \left\{ \langle (\dot{\Pi} + \Pi \widehat{A} + \widehat{A}^\top \Pi + \widehat{Q}) y, y \rangle \right.$$

$$+ 2 \langle (\Pi \widehat{B} + \widehat{S}^\top) \mathbb{E}[u], y \rangle + 2 \langle \dot{\bar{\eta}} + \widehat{A}^\top \bar{\eta} + \mathbb{E}[q] + \bar{q} + \Pi \mathbb{E}[b], y \rangle$$

$$\left. + 2 \langle \widehat{B}^\top \bar{\eta} + \mathbb{E}[\rho] + \bar{\rho}, \mathbb{E}[u] \rangle + \langle \widehat{R} \mathbb{E}[u], \mathbb{E}[u] \rangle + 2 \langle \bar{\eta}, \mathbb{E}[b] \rangle \right\} ds. \qquad (3.4.32)$$

Substituting (3.4.31) and (3.4.32) into (3.4.30) gives

$$J(t, \xi; u) = \mathbb{E} \langle P(t)(\xi - \mathbb{E}[\xi]) + 2\eta(t), \xi - \mathbb{E}[\xi] \rangle + \langle \Pi(t) \mathbb{E}[\xi] + 2\bar{\eta}(t), \mathbb{E}[\xi] \rangle$$

$$+ \mathbb{E} \int_t^T \left\{ \langle \mathcal{R}(P)(v - \Theta z - \varphi), v - \Theta z - \varphi \rangle - \langle \mathcal{R}(P)\varphi, \varphi \rangle \right.$$

$$\left. + \langle P\sigma, \sigma \rangle + 2 \langle \eta, b - \mathbb{E}[b] \rangle + 2 \langle \zeta, \sigma \rangle + 2 \langle \bar{\eta}, \mathbb{E}[b] \rangle \right\} ds$$

$$+ \int_t^T \left\{ \langle \widehat{\Theta}^\top \widehat{\mathcal{R}}(P) \widehat{\Theta} y, y \rangle - 2 \langle \widehat{\Theta}^\top \widehat{\mathcal{R}}(P) \mathbb{E}[u], y \rangle \right.$$

$$\left. + \langle \widehat{\mathcal{R}}(P) \mathbb{E}[u], \mathbb{E}[u] \rangle - 2 \langle \widehat{\mathcal{R}}(P) \bar{\varphi}, \mathbb{E}[u] - \widehat{\Theta} y \rangle \right\} ds$$

$$= \mathbb{E} \langle P(t)(\xi - \mathbb{E}[\xi]) + 2\eta(t), \xi - \mathbb{E}[\xi] \rangle + \langle \Pi(t) \mathbb{E}[\xi] + 2\bar{\eta}(t), \mathbb{E}[\xi] \rangle$$

$$+ \mathbb{E} \int_t^T \left\{ \langle P\sigma, \sigma \rangle + 2 \langle \eta, b - \mathbb{E}[b] \rangle + 2 \langle \zeta, \sigma \rangle + 2 \langle \bar{\eta}, \mathbb{E}[b] \rangle \right.$$

$$- \langle \widehat{\mathcal{R}}(P) \bar{\varphi}, \bar{\varphi} \rangle - \langle \mathcal{R}(P)(\varphi - \mathbb{E}[\varphi]), \varphi - \mathbb{E}[\varphi] \rangle$$

$$+ \langle \mathcal{R}(P)(v - \Theta z - \varphi + \mathbb{E}[\varphi]), v - \Theta z - \varphi + \mathbb{E}[\varphi] \rangle$$

$$\left. + \langle \widehat{\mathcal{R}}(P)(\mathbb{E}[u] - \widehat{\Theta} y - \bar{\varphi}), \mathbb{E}[u] - \widehat{\Theta} y - \bar{\varphi} \rangle \right\} ds. \qquad (3.4.33)$$

Since $\mathcal{R}(P), \widehat{\mathcal{R}}(P) \gg 0$, (3.4.33) implies that

$$J(t, \xi; u) \geqslant \mathbb{E} \langle P(t)(\xi - \mathbb{E}[\xi]) + 2\eta(t), \xi - \mathbb{E}[\xi] \rangle + \langle \Pi(t) \mathbb{E}[\xi] + 2\bar{\eta}(t), \mathbb{E}[\xi] \rangle$$

$$+ \mathbb{E} \int_t^T \left\{ \langle P\sigma, \sigma \rangle + 2 \langle \eta, b - \mathbb{E}[b] \rangle + 2 \langle \zeta, \sigma \rangle + 2 \langle \bar{\eta}, \mathbb{E}[b] \rangle \right.$$

$$\left. - \langle \mathcal{R}(P)(\varphi - \mathbb{E}[\varphi]), \varphi - \mathbb{E}[\varphi] \rangle - \langle \widehat{\mathcal{R}}(P) \bar{\varphi}, \bar{\varphi} \rangle \right\} ds, \qquad (3.4.34)$$

with the equality holding if and only if

$$v = \Theta z + \varphi - \mathbb{E}[\varphi], \quad \mathbb{E}[u] = \widehat{\Theta} y + \bar{\varphi},$$

or equivalently, if and only if

$$u = \Theta(X - \mathbb{E}[X]) + \widehat{\Theta}\mathbb{E}[X] + \varphi - \mathbb{E}[\varphi] + \bar{\varphi}. \qquad (3.4.35)$$

In particular, when $b, \sigma, g, \bar{g}, q, \bar{q}, \rho, \bar{\rho}$ vanish, we have

$$(\eta(s), \zeta(s)) = (0, 0), \quad \bar{\eta}(s) = 0, \quad \varphi(s) = \bar{\varphi}(s) = 0; \quad \forall t \leqslant s \leqslant T.$$

Then we may take $\xi = 0$ to get from (3.4.33) that

$$J^0(t, 0; u) = \mathbb{E}\int_t^T \Big\{ \langle \mathcal{R}(P)\{u - \mathbb{E}[u] - \Theta(X - \mathbb{E}[X])\}, u - \mathbb{E}[u] - \Theta(X - \mathbb{E}[X]) \rangle$$
$$+ \langle \widehat{\mathcal{R}}(P)(\mathbb{E}[u] - \widehat{\Theta}\mathbb{E}[X]), \mathbb{E}[u] - \widehat{\Theta}\mathbb{E}[X] \rangle \Big\} ds. \qquad (3.4.36)$$

Noting that $\mathcal{R}(P), \widehat{\mathcal{R}}(P) \geqslant \delta I$ for some $\delta > 0$ and using Lemma 3.4.5, we obtain

$$J^0(t, 0; u) \geqslant \delta\, \mathbb{E}\int_t^T \Big\{ |u - \mathbb{E}[u] - \Theta(X - \mathbb{E}[X])|^2 + |\mathbb{E}[u] - \widehat{\Theta}\mathbb{E}[X]|^2 \Big\} ds$$

$$\geqslant \delta\, \mathbb{E}\int_t^T \Big\{ |u - \Theta(X - \mathbb{E}[X])|^2 - 2\langle u - \Theta(X - \mathbb{E}[X]), \mathbb{E}[u] \rangle$$
$$+ (1 + \gamma)|\mathbb{E}[u]|^2 \Big\} ds$$

$$\geqslant \frac{\delta\gamma}{1 + \gamma}\, \mathbb{E}\int_t^T |u - \Theta(X - \mathbb{E}[X])|^2 ds$$

$$\geqslant \frac{\delta\gamma^2}{1 + \gamma}\, \mathbb{E}\int_t^T |u(s)|^2 ds, \quad \forall u \in \mathcal{U}[t, T],$$

for some $\gamma > 0$. The desired result follows. $\qquad \square$

To conclude this section, we present the following result, which is a direct consequence of (3.4.34)–(3.4.35).

Corollary 3.4.7 *Under the assumptions of Theorem 3.4.6, the unique optimal control of Problem (MFLQ) for the initial pair (t, ξ) is given by*

$$u^* = \Theta(X^* - \mathbb{E}[X^*]) + \widehat{\Theta}\mathbb{E}[X^*] + \varphi - \mathbb{E}[\varphi] + \bar{\varphi},$$

where $\Theta, \widehat{\Theta}$ are defined as in (3.4.28), $\varphi, \bar{\varphi}$ are defined as in (3.4.29), and X^* is the solution to the closed-loop system

$$\begin{cases} dX^*(s) = \big\{ (A + B\Theta)(X^* - \mathbb{E}[X^*]) + (\widehat{A} + \widehat{B}\widehat{\Theta})\mathbb{E}[X^*] \\ \qquad\qquad + B(\varphi - \mathbb{E}[\varphi]) + \widehat{B}\bar{\varphi} + b \big\} ds \\ \qquad\qquad + \big\{ (C + D\Theta)(X^* - \mathbb{E}[X^*]) + (\widehat{C} + \widehat{D}\widehat{\Theta})\mathbb{E}[X^*] \\ \qquad\qquad + D(\varphi - \mathbb{E}[\varphi]) + \widehat{D}\bar{\varphi} + \sigma \big\} dW, \quad s \in [t, T], \\ X^*(t) = \xi. \end{cases}$$

Moreover, the value $V(t, \xi)$ is given by

$$V(t, \xi) = \mathbb{E}\langle P(t)(\xi - \mathbb{E}[\xi]) + 2\eta(t), \xi - \mathbb{E}[\xi]\rangle + \langle \Pi(t)\mathbb{E}[\xi] + 2\bar{\eta}(t), \mathbb{E}[\xi]\rangle$$

$$+ \mathbb{E}\int_t^T \left\{ \langle P\sigma, \sigma\rangle + 2\langle \eta, b - \mathbb{E}[b]\rangle + 2\langle \zeta, \sigma\rangle + 2\langle \bar{\eta}, \mathbb{E}[b]\rangle \right.$$

$$\left. - \langle \mathcal{R}(P)(\varphi - \mathbb{E}[\varphi]), \varphi - \mathbb{E}[\varphi]\rangle - \langle \widehat{\mathcal{R}}(P)\bar{\varphi}, \bar{\varphi}\rangle \right\} ds.$$

In particular, the value function of Problem $(MFLQ)^0$ is given by

$$V^0(t, \xi) = \mathbb{E}\langle P(t)(\xi - \mathbb{E}[\xi]), \xi - \mathbb{E}[\xi]\rangle + \langle \Pi(t)\mathbb{E}[\xi], \mathbb{E}[\xi]\rangle.$$

3.4.3   Sufficient Conditions for the Uniform Convexity

In this section we present a couple of sufficient conditions for the uniform convexity of the cost functional.

Proposition 3.4.8 *Let (A1)–(A2) hold. If there exists a constant $\delta > 0$ such that for almost every $s \in [t, T]$,*

$$R(s), \ \widehat{R}(s) \geqslant \delta I_m, \qquad Q(s) - S(s)^\top R(s)^{-1} S(s) \geqslant 0,$$

$$G, \ \widehat{G} \geqslant 0, \qquad \widehat{Q}(s) - \widehat{S}(s)^\top \widehat{R}(s)^{-1} \widehat{S}(s) \geqslant 0,$$

then the uniform convexity condition (A4) holds.

Proof Denote by $X_0^u = \{X_0^u(s); t \leqslant s \leqslant T\}$ the solution of (3.4.24) corresponding to the control $u \in \mathcal{U}[t, T]$ and set

$$y(s) = X_0^u(s) - \mathbb{E}[X_0^u(s)], \quad v(s) = u(s) - \mathbb{E}[u(s)].$$

Then

$$J^0(t, 0; u) = \mathbb{E}\left\{ \langle Gy(T), y(T)\rangle + \int_t^T \left\langle \begin{pmatrix} Q & S^\top \\ S & R \end{pmatrix} \begin{pmatrix} y \\ v \end{pmatrix}, \begin{pmatrix} y \\ v \end{pmatrix} \right\rangle ds \right\}$$

$$+ \langle \widehat{G}\mathbb{E}[X_0^u(T)], \mathbb{E}[X_0^u(T)]\rangle$$

$$+ \int_t^T \left\langle \begin{pmatrix} \widehat{Q} & \widehat{S}^\top \\ \widehat{S} & \widehat{R} \end{pmatrix} \begin{pmatrix} \mathbb{E}[X_0^u] \\ \mathbb{E}[u] \end{pmatrix}, \begin{pmatrix} \mathbb{E}[X_0^u] \\ \mathbb{E}[u] \end{pmatrix} \right\rangle ds$$

$$\geqslant \int_t^T \left\langle \begin{pmatrix} Q & S^\top \\ S & R \end{pmatrix} \begin{pmatrix} y \\ v \end{pmatrix}, \begin{pmatrix} y \\ v \end{pmatrix} \right\rangle ds$$

$$+ \int_t^T \left\langle \begin{pmatrix} \widehat{Q} & \widehat{S}^\top \\ \widehat{S} & \widehat{R} \end{pmatrix} \begin{pmatrix} \mathbb{E}[X_0^u] \\ \mathbb{E}[u] \end{pmatrix}, \begin{pmatrix} \mathbb{E}[X_0^u] \\ \mathbb{E}[u] \end{pmatrix} \right\rangle ds. \qquad (3.4.37)$$

The first term on the right-hand side of (3.4.37) is equal to

$$\mathbb{E} \int_t^T \Big[\langle Qy, y \rangle + 2\langle Sy, v \rangle + \langle Rv, v \rangle \Big] ds$$

$$= \mathbb{E} \int_t^T \Big[\langle (Q - S^\top R^{-1} S) y, y \rangle + \langle R(v + R^{-1} Sy), v + R^{-1} Sy \rangle \Big] ds$$

$$\geqslant \delta \, \mathbb{E} \int_t^T |v + R^{-1} Sy|^2 \, ds.$$

Similarly, the second term on the right-hand side of (3.4.37) is equal to

$$\int_t^T \Big\{ \langle \widehat{Q} \mathbb{E}[X_0^u], \mathbb{E}[X_0^u] \rangle + 2\langle \widehat{S} \mathbb{E}[X_0^u], \mathbb{E}[u] \rangle + \langle \widehat{R} \mathbb{E}[u], \mathbb{E}[u] \rangle \Big\} ds$$

$$\geqslant \delta \int_t^T |\mathbb{E}[u] + \widehat{R}^{-1} \widehat{S} \mathbb{E}[X_0^u]|^2 \, ds.$$

Consequently, by using Lemma 3.4.5 with $\Theta = -R^{-1} S$ and $\bar{\Theta} = -\widehat{R}^{-1} \widehat{S}$, we obtain

$$J^0(t, 0; u) \geqslant \delta \, \mathbb{E} \int_t^T \Big\{ |u - \mathbb{E}[u] + R^{-1} S(X_0^u - \mathbb{E}[X_0^u])|^2 + \gamma |\mathbb{E}[u]|^2 \Big\} ds$$

$$\geqslant \frac{\delta \gamma}{1 + \gamma} \mathbb{E} \int_t^T |u + R^{-1} S(X_0^u - \mathbb{E}[X_0^u])|^2 \, ds$$

$$\geqslant \frac{\delta \gamma^2}{1 + \gamma} \mathbb{E} \int_t^T |u(s)|^2 ds, \quad \forall u \in \mathcal{U}[t, T].$$

This completes the proof. $\qquad\qquad\qquad\qquad\qquad\qquad\qquad\qquad\square$

Proposition 3.4.9 *Let (A1)–(A2) hold. Let P and Π be the solutions to the following Lyapunov equations for some $Q_0, \widehat{Q}_0 \in L^1(t, T; \bar{\mathbb{S}}_+^n)$, respectively:*

$$\begin{cases} \dot{P} + PA + A^\top P + C^\top PC + Q - Q_0 = 0, & s \in [t, T], \\ P(T) = G, \end{cases}$$

$$\begin{cases} \dot{\Pi} + \Pi \widehat{A} + \widehat{A}^\top \Pi + \widehat{C}^\top P \widehat{C} + \widehat{Q} - \widehat{Q}_0 = 0, & s \in [t, T], \\ \Pi(T) = \widehat{G}. \end{cases}$$

If for some $\delta > 0$,

$$\begin{pmatrix} Q_0 & \mathcal{S}(P)^\top \\ \mathcal{S}(P) & \mathcal{R}(P) - \delta I_m \end{pmatrix} \geqslant 0, \quad \begin{pmatrix} \widehat{Q}_0 & \widehat{\mathcal{S}}(P, \Pi)^\top \\ \widehat{\mathcal{S}}(P, \Pi) & \widehat{\mathcal{R}}(P) \end{pmatrix} \geqslant 0, \qquad (3.4.38)$$

a.e. on $[t, T]$, then the uniform convexity condition (A4) holds.

Proof Let $u \in \mathcal{U}[t, T]$ and X be the solution to the homogeneous state equation with initial state $\xi = 0$:

$$\begin{cases} dX(s) = \{AX + \bar{A}\mathbb{E}[X] + Bu + \bar{B}\mathbb{E}[u]\}ds \\ \qquad\qquad + \{CX + \bar{C}\mathbb{E}[X] + Du + \bar{D}\mathbb{E}[u]\}dW, \quad s \in [t, T], \\ X(t) = 0. \end{cases}$$

By letting

$$y(s) = \mathbb{E}[X(s)], \quad z(s) = X(s) - \mathbb{E}[X(s)], \quad v(s) = u(s) - \mathbb{E}[u(s)],$$

we can rewrite $J^0(t, 0; u)$ as follows:

$$\begin{aligned} J^0(t, 0; u) = \mathbb{E}\Bigg\{ \langle Gz(T), z(T) \rangle &+ \int_t^T \left\langle \begin{pmatrix} Q & S^\top \\ S & R \end{pmatrix} \begin{pmatrix} z \\ v \end{pmatrix}, \begin{pmatrix} z \\ v \end{pmatrix} \right\rangle ds \Bigg\} \\ &+ \langle \widehat{G}y(T), y(T) \rangle + \int_t^T \left\langle \begin{pmatrix} \widehat{Q} & \widehat{S}^\top \\ \widehat{S} & \widehat{R} \end{pmatrix} \begin{pmatrix} y \\ \mathbb{E}[u] \end{pmatrix}, \begin{pmatrix} y \\ \mathbb{E}[u] \end{pmatrix} \right\rangle ds. \end{aligned}$$

Observe that y and z satisfy, respectively, the following ODE and SDE:

$$\begin{cases} \dot{y}(s) = \widehat{A}y + \widehat{B}\mathbb{E}[u], \quad s \in [t, T], \\ y(t) = 0, \end{cases}$$

$$\begin{cases} dz(s) = (Az + Bv)ds + (Cz + Dv + \widehat{C}y + \widehat{D}\mathbb{E}[u])dW, \quad s \in [t, T], \\ z(t) = 0, \end{cases}$$

and that $\mathbb{E}[z] = 0$, $\mathbb{E}[v] = 0$. Then using integration by parts, we have

$$\begin{aligned} \mathbb{E}\langle Gz(T), z(T) \rangle = \mathbb{E} \int_t^T \Big\{ & \langle (\dot{P} + PA + A^\top P + C^\top PC)z, z \rangle \\ &+ 2\langle (PB + C^\top PD)v, z \rangle + \langle D^\top PDv, v \rangle \\ &+ \langle P(\widehat{C}y + \widehat{D}\mathbb{E}[u]), \widehat{C}y + \widehat{D}\mathbb{E}[u] \rangle \Big\} ds \\ = \mathbb{E} \int_t^T \Big\{ & \langle (Q_0 - Q)z, z \rangle + 2\langle (PB + C^\top PD)v, z \rangle \\ &+ \langle D^\top PDv, v \rangle + \langle P(\widehat{C}y + \widehat{D}\mathbb{E}[u]), \widehat{C}y + \widehat{D}\mathbb{E}[u] \rangle \Big\} ds. \end{aligned}$$

Similarly,

$$\langle \widehat{G}y(T), y(T) \rangle = \int_t^T \Big\{ \langle (\widehat{Q}_0 - \widehat{Q} - \widehat{C}^\top P\widehat{C})y, y \rangle + 2\langle \Pi \widehat{B}\mathbb{E}[u], y \rangle \Big\} ds.$$

Consequently,

$$
J^0(t, 0; u) = \mathbb{E} \int_t^T \left\langle \begin{pmatrix} Q_0 & S(P)^\top \\ S(P) & R(P) \end{pmatrix} \begin{pmatrix} z \\ v \end{pmatrix}, \begin{pmatrix} z \\ v \end{pmatrix} \right\rangle ds
$$
$$
+ \int_t^T \left\langle \begin{pmatrix} \widehat{Q}_0 & \widehat{S}(P, \Pi)^\top \\ \widehat{S}(P, \Pi) & \widehat{R}(P) \end{pmatrix} \begin{pmatrix} y \\ \mathbb{E}[u] \end{pmatrix}, \begin{pmatrix} y \\ \mathbb{E}[u] \end{pmatrix} \right\rangle ds.
$$

The desired result follows from (3.4.38). $\qquad\square$

3.5 Multi-dimensional Brownian Motion Case and Applications

In the previous sections, the Brownian motion has been assumed to be one-dimensional for the purpose of simplicity. In the case of a d-dimensional Brownian motion $W = \{W(t) = (W_1(t), \ldots, W_d(t))^\top; 0 \leqslant t < \infty\}$, the MFLQ problem is to find a control $u^* \in \mathcal{U}[t, T]$ such that the quadratic cost functional (3.1.2) is minimized subject to the following state equation over $[t, T]$:

$$
\begin{cases}
dX(s) = \{AX + \bar{A}\mathbb{E}[X] + Bu + \bar{B}\mathbb{E}[u] + b\}ds \\
\qquad\quad + \displaystyle\sum_{i=1}^{d} \{C_i X + \bar{C}_i \mathbb{E}[X] + D_i u + \bar{D}_i \mathbb{E}[u] + \sigma_i\}dW_i, \qquad (3.5.1) \\
X(t) = \xi,
\end{cases}
$$

where the weighting matrices in the cost functional satisfy (A2), and the coefficients of the state equation satisfy the following assumption that is similar to (A1):

(A1)* The coefficients of (3.5.1) satisfy

$$
\begin{cases}
A, \bar{A} \in L^1(0, T; \mathbb{R}^{n \times n}), & B, \bar{B} \in L^2(0, T; \mathbb{R}^{n \times m}), \\
C_i, \bar{C}_i \in L^2(0, T; \mathbb{R}^{n \times n}), & D_i, \bar{D}_i \in L^\infty(0, T; \mathbb{R}^{n \times m}), \\
b \in L^2_{\mathbb{F}}(\Omega; L^1(0, T; \mathbb{R}^n)), & \sigma_i \in L^2_{\mathbb{F}}(0, T; \mathbb{R}^n); \quad i = 1, \ldots, d.
\end{cases}
$$

In this case, the associated Riccati equations (3.4.1) and (3.4.2) become

$$\begin{cases} \dot{P} + PA + A^\top P + \sum_{i=1}^{d} C_i^\top P C_i + Q - \left(PB + \sum_{i=1}^{d} C_i^\top P D_i + S^\top \right) \\ \quad \times \left(R + \sum_{i=1}^{d} D_i^\top P D_i \right)^{-1} \left(B^\top P + \sum_{i=1}^{d} D_i^\top P C_i + S \right) = 0, \\ P(T) = G, \end{cases} \tag{3.5.2}$$

and

$$\begin{cases} \dot{\Pi} + \Pi\widehat{A} + \widehat{A}^\top \Pi + \widehat{Q} + \sum_{i=1}^{d} \widehat{C}_i^\top P \widehat{C}_i - \left(\Pi\widehat{B} + \sum_{i=1}^{d} \widehat{C}_i^\top P \widehat{D}_i + \widehat{S}^\top \right) \\ \quad \times \left(\widehat{R} + \sum_{i=1}^{d} \widehat{D}_i^\top P \widehat{D}_i \right)^{-1} \left(\widehat{B}^\top \Pi + \sum_{i=1}^{d} \widehat{D}_i^\top P \widehat{C}_i + \widehat{S} \right) = 0, \\ \Pi(T) = \widehat{G}, \end{cases} \tag{3.5.3}$$

respectively, where $\widehat{A}, \widehat{B}, \widehat{Q}, \widehat{S}, \widehat{R}, \widehat{G}$ are as in (3.4.4) and

$$\widehat{C}_i = C_i + \bar{C}_i, \quad \widehat{D}_i = D_i + \bar{D}_i; \quad 1 \leqslant i \leqslant d.$$

Similar to the case of one-dimensional Brownian motion, we have the following result.

Theorem 3.5.1 *Let (A1)* and (A2) hold. The uniform convexity condition (A4) holds if and only if*

(i) the Riccati equation (3.5.2) admits a solution $P \in C([t, T]; \mathbb{S}^n)$ such that

$$\mathcal{R}(P) \triangleq R + \sum_{i=1}^{d} D_i^\top P D_i \gg 0, \quad \widehat{\mathcal{R}}(P) \triangleq \widehat{R} + \sum_{i=1}^{d} \widehat{D}_i^\top P \widehat{D}_i \gg 0,$$

(ii) the Riccati equation (3.5.3) admits a solution $\Pi \in C([t, T]; \mathbb{S}^n)$.

In this case, the unique optimal control of Problem (MFLQ) for the initial pair (t, ξ) is given by

$$u^* = \Theta(X^* - \mathbb{E}[X^*]) + \widehat{\Theta}\mathbb{E}[X^*] + \varphi - \mathbb{E}[\varphi] + \bar{\varphi},$$

where $\Theta, \widehat{\Theta}, \varphi, \bar{\varphi}$ are defined by

$$\Theta \triangleq -\mathcal{R}(P)^{-1}\left(B^{\top}P + \sum_{i=1}^{d} D_i^{\top}PC_i + S\right),$$

$$\widehat{\Theta} \triangleq -\widehat{\mathcal{R}}(P)^{-1}\left(\widehat{B}^{\top}\Pi + \sum_{i=1}^{d} \widehat{D}_i^{\top}P\widehat{C}_i + \widehat{S}\right),$$

$$\varphi \triangleq -\mathcal{R}(P)^{-1}\left\{B^{\top}\eta + \sum_{i=1}^{d} D_i^{\top}(\zeta_i + P\sigma_i) + \rho\right\},$$

$$\bar{\varphi} \triangleq -\widehat{\mathcal{R}}(P)^{-1}\left\{\widehat{B}^{\top}\bar{\eta} + \sum_{i=1}^{d} \widehat{D}_i^{\top}(\mathbb{E}[\zeta_i] + P\mathbb{E}[\sigma_i]) + \mathbb{E}[\rho] + \bar{\rho}\right\},$$

with (η, ζ) and $\bar{\eta}$ being the solutions to the BSDE

$$\begin{cases} d\eta(s) = -\Big[(A + B\Theta)^{\top}\eta + \sum_{i=1}^{d}(C_i + D_i\Theta)^{\top}(\zeta_i + P\sigma_i) \\ \qquad\qquad + \Theta^{\top}\rho + Pb + q\Big]ds + \sum_{i=1}^{d}\zeta_i dW_i, \quad s \in [t, T], \\ \eta(T) = g, \end{cases} \tag{3.5.4}$$

and the ODE

$$\begin{cases} \dot{\bar{\eta}} + (\widehat{A} + \widehat{B}\widehat{\Theta})^{\top}\bar{\eta} + \widehat{\Theta}^{\top}\Big\{\sum_{i=1}^{d}\widehat{D}_i^{\top}(P\mathbb{E}[\sigma_i] + \mathbb{E}[\zeta_i]) + \mathbb{E}[\rho] + \bar{\rho}\Big\} \\ \qquad + \sum_{i=1}^{d}\widehat{C}_i^{\top}(P\mathbb{E}[\sigma_i] + \mathbb{E}[\zeta_i]) + \mathbb{E}[q] + \bar{q} + \Pi\mathbb{E}[b] = 0, \quad s \in [t, T], \\ \bar{\eta}(T) = \mathbb{E}[g] + \bar{g}, \end{cases} \tag{3.5.5}$$

respectively, and X^* is the solution to the closed-loop system

$$\begin{cases} dX^*(s) = \Big\{(A + B\Theta)(X^* - \mathbb{E}[X^*]) + (\widehat{A} + \widehat{B}\widehat{\Theta})\mathbb{E}[X^*] \\ \qquad\qquad + B(\varphi - \mathbb{E}[\varphi]) + \widehat{B}\bar{\varphi} + b\Big\}ds \\ \qquad\qquad + \displaystyle\sum_{i=1}^{d}\Big\{(C_i + D_i\Theta)(X^* - \mathbb{E}[X^*]) + (\widehat{C}_i + \widehat{D}_i\widehat{\Theta})\mathbb{E}[X^*] \\ \qquad\qquad + D_i(\varphi - \mathbb{E}[\varphi]) + \widehat{D}_i\bar{\varphi} + \sigma_i\Big\}dW_i, \quad s \in [t, T], \\ X^*(t) = \xi. \end{cases}$$

Moreover, the value $V(t, \xi)$ is given by

$$\begin{aligned} V(t, \xi) &= \mathbb{E}\langle P(t)(\xi - \mathbb{E}[\xi]) + 2\eta(t), \xi - \mathbb{E}[\xi]\rangle + \langle \Pi(t)\mathbb{E}[\xi] + 2\bar{\eta}(t), \mathbb{E}[\xi]\rangle \\ &\quad + \mathbb{E}\int_t^T \Big\{ \sum_{i=1}^{d}\langle P\sigma_i, \sigma_i\rangle + 2\langle \eta, b - \mathbb{E}[b]\rangle + \sum_{i=1}^{d} 2\langle \zeta_i, \sigma_i\rangle \\ &\quad + 2\langle\bar{\eta}, \mathbb{E}[b]\rangle - \langle \mathcal{R}(P)(\varphi - \mathbb{E}[\varphi]), \varphi - \mathbb{E}[\varphi]\rangle - \langle\widehat{\mathcal{R}}(P)\bar{\varphi}, \bar{\varphi}\rangle \Big\}ds. \end{aligned}$$

In particular, the value function of Problem (MFLQ)^0 is given by

$$V^0(t, \xi) = \mathbb{E}\langle P(t)(\xi - \mathbb{E}[\xi]), \xi - \mathbb{E}[\xi]\rangle + \langle \Pi(t)\mathbb{E}[\xi], \mathbb{E}[\xi]\rangle.$$

As an application of the mean-field control theory, let us consider a market in which $d + 1$ assets (or "securities") are traded continuously. One of the assets, called the *bond*, has a price $P_0(t)$ which evolves according to the ODE

$$\begin{cases} dP_0(t) = r(t)P_0(t)dt, \quad t \in [0, T], \\ P_0(0) = p_0. \end{cases} \tag{3.5.6}$$

The remaining d assets, called *stocks*, are "risky"; their prices are modeled by the linear SDE for $i = 1, \ldots, d$:

$$
\begin{cases}
dP_i(t) = P_i(t)b_i(t)dt + P_i(t) \sum_{j=1}^{d} \sigma_{ij}(t)dW_j(t), & t \in [0, T], \\
P_i(0) = p_i.
\end{cases}
\tag{3.5.7}
$$

The process $W = \{W(t) = (W_1(t), \ldots, W_d(t))^\top; 0 \leqslant t \leqslant T\}$ is a standard d-dimensional Brownian motion on a probability space $(\Omega, \mathcal{F}, \mathbb{P})$, and the filtration $\mathbb{F} = \{\mathcal{F}_t\}_{t \geqslant 0}$ is the augmentation under $\mathbb{P}$ of the natural filtration generated by W. The *interest rate* $r(t)$, as well as the vector of *mean rates of return* $b(t) = (b_1(t), \ldots, b_d(t))$ and the *dispersion* matrix $\sigma(t) = (\sigma_{ij}(t))_{1 \leqslant i, j \leqslant d}$, is assumed to be deterministic (nonrandom) and bounded uniformly in $t \in [0, T]$. We assume that for some number $\delta > 0$,

$$
\sigma(t)\sigma(t)^\top \geqslant \delta I_d, \quad \forall t \in [0, T].
\tag{3.5.8}
$$

We imagine now an investor who starts with some initial wealth $x_0 > 0$ and invests it in the $d + 1$ assets described previously. Let $N_i(t)$ denote the number of shares of asset i owned by the investor at time t. Then the investor's wealth at time t is

$$
x(t) = \sum_{i=0}^{d} N_i(t)P_i(t).
\tag{3.5.9}
$$

Assume that the trading of shares is self-financed and takes place continuously, and that transaction cost and consumptions are not considered. Then one has

$$
\begin{cases}
dx(t) = \sum_{i=0}^{d} N_i(t)dP_i(t), & t \in [0, T], \\
x(0) = x_0.
\end{cases}
\tag{3.5.10}
$$

Taking (3.5.6), (3.5.7), (3.5.9) into account and denoting by

$$
u_i(t) \triangleq N_i(t)P_i(t)
$$

the amount invested in the i-th stock, $1 < i < d$, we may write (3.5.10) as

$$
\begin{cases}
dx(t) = \left\{ r(t)x(t) + \sum_{i=1}^{d} [b_i(t) - r(t)]u_i(t) \right\} dt \\
\qquad + \sum_{i=1}^{d} \sum_{j=1}^{d} \sigma_{ij}(t)u_i(t)dW_j(t), \quad t \in [0, T], \\
x(0) = x_0.
\end{cases}
\tag{3.5.11}
$$

Definition 3.5.2 A *portfolio process* $u = \{u(t) = (u_1(t), \ldots, u_d(t))^\top, \mathcal{F}_t; 0 \leqslant t \leqslant T\}$ is a progressively measurable process for which

$$
\mathbb{E} \int_0^T |u(t)|^2 dt < \infty.
$$

In other words, a portfolio process is an element in $L_{\mathbb{F}}^2(0, T; \mathbb{R}^d)$.

The objective of the investor is to maximize the mean terminal wealth $\mathbb{E}x(T)$, or equivalently, to minimize

$$
h_1(u) \triangleq -\mathbb{E}x(T),
$$

and at the same time to minimize the variance of the terminal wealth

$$
h_2(u) \triangleq \operatorname{var} x(T) = \mathbb{E}x(T)^2 - [\mathbb{E}x(T)]^2.
$$

Definition 3.5.3 A portfolio process u^* is called an *efficient portfolio* if there exists no portfolio process u such that

$$
h_1(u) \leqslant h_1(u^*), \quad h_2(u) \leqslant h_2(u^*),
$$

and at least one of the inequalities is strict. In this case, $(h_1(u^*), h_2(u^*)) \in \mathbb{R}^2$ is called an *efficient point*.

In other words, an efficient portfolio is one where there exists no other portfolio better than it with respect to both the mean and variance criteria. The problem then is to identify the efficient portfolios and is referred to as a *mean-variance portfolio selection problem* (MV problem, for short).

An efficient portfolio can be found by solving a single-objective optimization problem where the objective is a weighted average of the two original criteria, as shown by the following result.

Proposition 3.5.4 *Let $\lambda > 0$ be a given real number. If u^* is a minimum of $\lambda h_1(u) + h_2(u)$, then it is an efficient portfolio of the MV problem.*

Proof We prove the conclusion by contradiction. Suppose that there exists a portfolio v such that, say,

$$h_1(v) < h_1(u^*), \quad h_2(v) \leqslant h_2(u^*).$$

Then $\lambda h_1(v) + h_2(v) < \lambda h_1(u^*) + h_2(u^*)$, which contradicts to the fact that u^* is a minimum of $\lambda h_1(u) + h_2(u)$. □

Let us now consider the following MFLQ problem for $\lambda > 0$.

Problem (MFLQ)$^\lambda$. For given initial state $x_0 \in \mathbb{R}$, find a control $u^* \in L_\mathbb{F}^2(0, T; \mathbb{R}^d)$ such that

$$J_\lambda(x_0; u^*) = \inf_{u \in L_\mathbb{F}^2(0,T;\mathbb{R}^d)} J_\lambda(x_0; u), \qquad (3.5.12)$$

subject to (3.5.11), where

$$J_\lambda(x_0; u) \triangleq 2\lambda h_1(u) + h_2(u) = \mathbb{E}x(T)^2 - [\mathbb{E}x(T)]^2 - 2\lambda \mathbb{E}x(T).$$

According to Proposition 3.5.4, we may find efficient portfolios of the MV problem by solving the above Problem (MFLQ)$^\lambda$, in which the initial time has been taken to be $t = 0$. We now apply Theorem 3.5.1 to get optimal control for Problem (MFLQ)$^\lambda$.

The Riccati equations associated to Problem (MFLQ)$^\lambda$ are

$$\begin{cases} \dot{P} + PA + A^\top P - PB\left(\sum_{i=1}^d D_i^\top P D_i\right)^{-1} B^\top P = 0, \\ P(T) = G, \end{cases} \qquad (3.5.13)$$

$$\begin{cases} \dot{\Pi} + \Pi\widehat{A} + \widehat{A}^\top \Pi - \Pi\widehat{B}\left(\sum_{i=1}^d \widehat{D}_i^\top P \widehat{D}_i\right)^{-1} \widehat{B}^\top \Pi = 0, \\ \Pi(T) = \widehat{G}, \end{cases} \qquad (3.5.14)$$

in which

$$A(t) = \widehat{A}(t) = r(t), \qquad B(t) = \widehat{B}(t) = (b_1(t) - r(t), \ldots, b_d(t) - r(t)),$$
$$G = 1, \quad \widehat{G} = 0, \qquad D_i(t) = \widehat{D}_i(t) = (\sigma_{1i}(t), \ldots, \sigma_{di}(t)).$$

Note that $P(t) \in \mathbb{R}$ and

$$\sum_{i=1}^d D_i^\top D_i = \sigma(t)\sigma(t)^\top \geqslant \delta I_d,$$

With the notation

$$\mu(t) = B(t)\left[\sigma(t)\sigma(t)^\top\right]^{-1} B(t)^\top,$$

we can rewrite (3.5.13) and (3.5.14) as

$$\begin{cases} \dot{P}(t) = [\mu(t) - 2r(t)]P(t), & t \in [0, T], \\ P(T) = 1, \end{cases} \tag{3.5.15}$$

$$\begin{cases} \dot{\Pi}(t) = \dfrac{\mu(t)}{P(t)}\Pi(t)^2 - 2r(t)\Pi(t), & t \in [0, T], \\ \Pi(T) = 0. \end{cases} \tag{3.5.16}$$

Clearly, the positive function

$$P(t) = e^{\int_t^T [2r(s) - \mu(s)]ds}, \quad 0 \leqslant t \leqslant T, \tag{3.5.17}$$

is the solution of (3.5.15) such that

$$\sum_{i=1}^d D_i^\top P D_i = \sum_{i=1}^d \widehat{D}_i^\top P \widehat{D}_i \gg 0,$$

and $\Pi(t) \equiv 0$ is the solution of (3.5.16). Observe that for Problem (MFLQ)$^\lambda$,

$$\Theta(t) = -\left[\sigma(t)\sigma(t)^\top\right]^{-1} B(t)^\top, \quad \widehat{\Theta}(t) = 0; \quad 0 \leqslant t \leqslant T,$$

and that all the coefficients in (3.5.4) are deterministic. Thus, the BSDE (3.5.4) reduces to the following ODE:

$$\begin{cases} d\eta(t) = -[r(t) - \mu(t)]\eta(t)dt, & t \in [0, T], \\ \eta(T) = 0, \end{cases}$$

and the ODE (3.5.5) reduces to

$$\begin{cases} \dot{\bar{\eta}}(t) + r(t)\bar{\eta}(t) = 0, & t \in [0, T], \\ \bar{\eta}(T) = -\lambda. \end{cases}$$

Clearly, the solutions of the above two ODEs are given by

$$\eta(t) = 0, \quad \bar{\eta}(t) = -\lambda e^{\int_t^T r(s)ds}; \quad 0 \leqslant t \leqslant T.$$

Therefore,

$$\varphi(t) = 0, \quad \bar{\varphi}(t) = -\lambda\Theta(t)e^{\int_t^T [\mu(s) - r(s)]ds}; \quad 0 \leqslant t \leqslant T,$$

and the optimal control of Problem (MFLQ)$^\lambda$ is given by

$$u_\lambda^*(t) = -\Theta(t)\left\{\lambda e^{\int_t^T [\mu(s)-r(s)]ds} - [x^*(t) - \mathbb{E}x^*(t)]\right\}. \tag{3.5.18}$$

Note that $\mathbb{E}x^*(t)$ satisfies the following ODE:

$$\begin{cases} dy(t) = \left[r(t)y(t) + B(t)\mathbb{E}u_\lambda^*(t)\right]dt \\ \qquad = \left[r(t)y(t) + \lambda\mu(t)e^{\int_t^T [\mu(s)-r(s)]ds}\right]dt, \quad t \in [0, T], \\ y(0) = x_0, \end{cases}$$

from which we get

$$\begin{aligned} \mathbb{E}x^*(t) &= x_0 e^{\int_0^t r(s)ds} + \lambda e^{-\int_t^T r(s)ds}\int_0^t \mu(s)e^{\int_s^T \mu(v)dv}ds \\ &= x_0 e^{\int_0^t r(s)ds} + \lambda e^{-\int_t^T r(s)ds}\left[e^{\int_0^T \mu(s)ds} - e^{\int_t^T \mu(s)ds}\right]. \end{aligned}$$

So (3.5.18) can be further written as

$$u_\lambda^*(t) = -\Theta(t)\left[x_0 e^{\int_0^t r(s)ds} + \lambda e^{\int_0^T \mu(s)ds - \int_t^T r(s)ds} - x^*(t)\right]. \tag{3.5.19}$$

Moreover, from the last assertion of Theorem 3.5.1 we have

$$\begin{aligned} J_\lambda(x_0; u_\lambda^*) &= 2\langle \bar{\eta}(0), x_0\rangle - \int_0^T \left\langle \left[\sigma(t)\sigma(t)^\top\right]P(t)\bar{\varphi}(t), \bar{\varphi}(t)\right\rangle dt \\ &= -2\lambda x_0 e^{\int_0^T r(s)ds} - \lambda^2 \int_0^T P(t)\mu(t)e^{\int_t^T 2[\mu(s)-r(s)]ds}dt. \end{aligned}$$

Substituting (3.5.17) into the above gives

$$\begin{aligned} J_\lambda(x_0; u_\lambda^*) &= -2\lambda x_0 e^{\int_0^T r(s)ds} - \lambda^2 \int_0^T \mu(t)e^{\int_t^T \mu(s)ds}dt \\ &= -2\lambda x_0 e^{\int_0^T r(s)ds} - \lambda^2\left[e^{\int_0^T \mu(s)ds} - 1\right]. \end{aligned}$$

Therefore, the efficient point corresponding to the efficient portfolio (3.5.19) is given by

$$\mathbb{E}x^*(T) = x_0 e^{\int_0^T r(s)ds} + \lambda\left[e^{\int_0^T \mu(s)ds} - 1\right],$$

$$\text{var}\, x^*(T) = J_\lambda(x_0; u_\lambda^*) + 2\lambda\mathbb{E}x^*(T) = \lambda^2\left[e^{\int_0^T \mu(s)ds} - 1\right].$$

3.6   Problems in Infinite Horizons

In this section, we consider mean-field linear-quadratic optimal control problems over the infinite time horizon $[0, \infty)$. For simplicity of presentation, we consider only the homogeneous problem. Thus, the state equation of interest here takes the following form:

$$\begin{cases} dX(s) = \{AX(s) + \bar{A}\mathbb{E}[X(s)] + Bu(s) + \bar{B}\mathbb{E}[u(s)]\}ds \\ \qquad\qquad + \{CX(s) + \bar{C}\mathbb{E}[X(s)] + Du(s) + \bar{D}\mathbb{E}[u(s)]\}dW, \qquad (3.6.1) \\ X(0) = x, \end{cases}$$

and the cost functional is of the form

$$J(x; u) = \mathbb{E}\int_0^\infty \Big\{ \langle QX(s), X(s)\rangle + \langle \bar{Q}\mathbb{E}[X(s)], \mathbb{E}[X(s)]\rangle$$
$$+ \langle Ru(s), u(s)\rangle + \langle \bar{R}\mathbb{E}[u(s)], \mathbb{E}[u(s)]\rangle \Big\}ds. \qquad (3.6.2)$$

In the above, the coefficients of (3.6.1) and the weighting matrices in (3.6.2) satisfy the following two assumptions, respectively.

(A1)′ $A, \bar{A}, C, \bar{C} \in \mathbb{R}^{n\times n}$, $B, \bar{B}, D, \bar{D} \in \mathbb{R}^{n\times m}$ are constant-valued.

(A2)′ $Q, \bar{Q} \in \mathbb{S}^n$, $R, \bar{R} \in \mathbb{S}^m$ are constant-valued and

$$Q, \ Q + \bar{Q} \geqslant 0, \quad R, \ R + \bar{R} > 0.$$

For convenience, let us recall the following notation:

$$L_{\mathbb{F}}^2(\mathbb{R}^n) = \Big\{ \varphi : [0, \infty) \times \Omega \to \mathbb{R}^n \mid \varphi \in \mathbb{F} \text{ and } \mathbb{E}\int_0^\infty |\varphi(t)|^2 dt < \infty \Big\},$$

$$\mathcal{X}_{loc}[0, \infty) = \Big\{ \varphi : [0, \infty) \times \Omega \to \mathbb{R}^n \mid \varphi \text{ is } \mathbb{F}\text{-adapted, continuous,}$$
$$\text{and } \mathbb{E}\Big[\sup_{0\leqslant t\leqslant T} |\varphi(t)|^2 \Big] < \infty \text{ for every } T > 0 \Big\},$$

$$\mathcal{X}[0, \infty) = \Big\{ \varphi \in \mathcal{X}_{loc}[0, \infty) \mid \mathbb{E}\int_0^\infty |\varphi(t)|^2 dt < \infty \Big\}.$$

By a standard argument using contraction mapping theorem, one can show that for any $(x, u) \in \mathbb{R}^n \times L_{\mathbb{F}}^2(\mathbb{R}^n)$, the state Equation (3.6.1) admits a unique solution $X(\cdot) = X(\cdot\,; x, u) \in \mathcal{X}_{loc}[0, \infty)$. It should be noted that in general, the solution X of (3.6.1) might not be in $\mathcal{X}[0, \infty)$ and thereby the cost functional $J(x; u)$ might not be defined. So we introduce the following set:

$$\mathcal{U}_{ad}(x) = \big\{ u \in L_{\mathbb{F}}^2(\mathbb{R}^m) \mid J(x; u) \text{ is defined} \big\}.$$

An element $u \in \mathcal{U}_{ad}(x)$ is called an *admissible control* for the initial state x, and the corresponding state process $X(\cdot) \equiv X(\cdot\,; x, u)$ is called an *admissible state process*. Our optimal control problem can then be stated as follows.

Problem (MFLQ)$_\infty$. For given $x \in \mathbb{R}^n$, find an admissible control $u^* \in \mathcal{U}_{ad}(x)$ such that

$$J(x; u^*) = \inf_{u \in \mathcal{U}_{ad}(x)} J(x; u) \equiv V(x).$$

The structure of $\mathcal{U}_{ad}(x)$ is kind of complicated, since it involves not only the state equation, but also the cost functional. Also, observe that the cost functional $J(x; u)$ can be written as

$$J(x; u) = \mathbb{E} \int_0^\infty \Big(|Q^{\frac{1}{2}}\{X(s) - \mathbb{E}[X(s)]\}|^2 + |(Q + \bar{Q})^{\frac{1}{2}}\mathbb{E}[X(s)]|^2$$
$$+ \langle Ru(s), u(s)\rangle + \langle \bar{R}\mathbb{E}[u(s)], \mathbb{E}[u(s)]\rangle \Big) ds.$$

So $u \in \mathcal{U}_{ad}(x)$ if and only if

$$\mathbb{E} \int_0^\infty \Big(|Q^{\frac{1}{2}}\{X(s) - \mathbb{E}[X(s)]\}|^2 + |(Q + \bar{Q})^{\frac{1}{2}}\mathbb{E}[X(s)]|^2 \Big) ds < \infty. \qquad (3.6.3)$$

Since Q and/or $(Q + \bar{Q})$ might be degenerate, we might not have $X \in \mathcal{X}[0, \infty)$ even though $u \in \mathcal{U}_{ad}(x)$. Later, we will give some better description of $\mathcal{U}_{ad}(x)$ under certain conditions.

3.6.1 Stability and Stabilizability

Let us first look at an example, in which the admissible control set $\mathcal{U}_{ad}(x)$ is empty for nonzero initial state x.

Example 3.6.1 Consider the one-dimensional controlled system

$$dX(s) = X(s)ds + \{\mathbb{E}[X(s)] + u(s)\}dW, \quad s \geq 0,$$

and the cost functional

$$J(x; u) = \mathbb{E} \int_0^\infty \Big\{|X(s)|^2 + |u(s)|^2 + |\mathbb{E}[u(s)]|^2\Big\} ds.$$

Clearly,

$$d\mathbb{E}[X(s)] = \mathbb{E}[X(s)]ds, \quad s \geq 0,$$

so with initial state x, $\mathbb{E}[X(s)] = xe^s$. Consequently,

$$dX(s) = X(s)ds + [xe^s + u(s)]dW, \quad s \geqslant 0,$$

from which we get

$$X(t) = e^t \left\{ x + \int_0^t [x + e^{-s}u(s)]dW \right\}, \quad t \geqslant 0.$$

As long as $x \neq 0$,

$$J(x; u) \geqslant \mathbb{E} \int_0^\infty |X(s)|^2 ds = \int_0^\infty e^{2t} \left\{ x^2 + \int_0^t [x + e^{-s}u(s)]^2 ds \right\} dt = \infty.$$

In this case, $\mathcal{U}_{ad}(x) = \varnothing$. Therefore, the corresponding Problem (MFLQ)$_\infty$ is not meaningful.

From the above example, we see that before investigating Problem (MFLQ)$_\infty$, one should find conditions for the system and the cost functional so that the set $\mathcal{U}_{ad}(x)$ is at least non-empty and hopefully it admits an accessible characterization. To this end, we first consider the following uncontrolled linear mean-field SDE (which amounts to taking $u = 0$ or letting $B = \bar{B} = D = \bar{D} = 0$):

$$\begin{cases} dX(t) = \{AX(t) + \bar{A}\mathbb{E}[X(t)]\}dt + \{CX(t) + \bar{C}\mathbb{E}[X(t)]\}dW, \\ X(0) = x. \end{cases} \tag{3.6.4}$$

We will briefly denote the above system by $[A, \bar{A}, C, \bar{C}]$, and for simplicity, we write $[A, C] = [A, 0, C, 0]$ when the mean-field terms are absent.

Definition 3.6.2 Let (A1)′–(A2)′ hold. The system $[A, \bar{A}, C, \bar{C}]$ is said to be

(i) L^2-*exponentially stable* if for every initial state $x \in \mathbb{R}^n$, the solution X of (3.6.4) satisfies

$$\lim_{t \to \infty} e^{\lambda t} \mathbb{E}|X(t)|^2 = 0$$

for some $\lambda > 0$;

(ii) L^2-*globally integrable* if for every $x \in \mathbb{R}^n$, the solution X of (3.6.4) is in $\mathcal{X}[0, \infty)$, i.e.,

$$\int_0^\infty \mathbb{E}|X(t)|^2 dt < \infty;$$

(iii) L^2-*asymptotically stable* if for every $x \in \mathbb{R}^n$, the solution X of (3.6.4) satisfies

$$\lim_{t \to \infty} \mathbb{E}|X(t)|^2 = 0; \tag{3.6.5}$$

(iv) $L^2_{Q,\bar{Q}}$-*globally integrable* if for every $x \in \mathbb{R}^n$, the solution X of (3.6.4) satisfies (3.6.3).

We shall simply say that the system $[A, \bar{A}, C, \bar{C}]$ is L^2_Q-*globally integrable* if it is $L^2_{Q,0}$-globally integrable.

Remark 3.6.3 We shall say that a matrix $M \in \mathbb{R}^{n \times n}$ is *exponentially stable* if the system $[M, 0, 0, 0]$ is L^2-exponentially stable.

Proposition 3.6.4 *Let (A1)′–(A2)′ hold. Among the statements*

  (i) *the system $[A, \bar{A}, C, \bar{C}]$ is L^2-exponentially stable;*
 (ii) *the system $[A, \bar{A}, C, \bar{C}]$ is L^2-globally integrable;*
(iii) *the system $[A, \bar{A}, C, \bar{C}]$ is L^2-asymptotically stable;*
 (iv) *the system $[A, \bar{A}, C, \bar{C}]$ is $L^2_{Q,\bar{Q}}$-globally integrable,*

we have the following implications:

$$(i) \implies (ii) \implies (iii); \qquad (ii) \implies (iv).$$

Proof The implications (i) $\Rightarrow$ (ii) and (ii) $\Rightarrow$ (iv) are trivial. For the implication (ii) $\Rightarrow$ (iii), we note that

$$\mathbb{E}|X(t)|^2 - \mathbb{E}|X(s)|^2$$
$$= \mathbb{E} \int_s^t \left\{ 2\langle X(s), AX(s) + \bar{A}\mathbb{E}[X(s)]\rangle + |CX(s) + \bar{C}\mathbb{E}[X(s)]|^2 \right\} ds$$
$$\leqslant L \int_s^t \mathbb{E}|X(s)|^2 ds. \tag{3.6.6}$$

Here and hereafter, $L > 0$ stands for a generic constant which could be different from line to line. Since

$$\int_s^t \mathbb{E}|X(s)|^2 ds \leqslant \mathbb{E} \int_0^\infty |X(s)|^2 ds < \infty,$$

(3.6.6) implies that $\mathbb{E}|X(t)|^2$ is bounded uniformly in $t \in [0, \infty)$ and thereby Lipschitz-continuous. Combining the Lipschitz-continuity and the integrability of $\mathbb{E}|X(t)|^2$ over $[0, \infty)$, we obtain (3.6.5). $\qquad \square$

Theorem 3.6.5 *Let (A1)′ hold.*

 (i) *If the system $[A, \bar{A}, C, \bar{C}]$ is L^2-asymptotically stable, then $A + \bar{A}$ is exponentially stable.*
(ii) *If $A + \bar{A}$ is exponentially stable, then the system $[A, \bar{A}, C, \bar{C}]$ is L^2-exponentially stable if either $[A, C]$ is L^2-globally integrable or*

$$C + \bar{C} = 0. \tag{3.6.7}$$

Proof (i) Suppose that (3.6.5) holds. Taking expectations in (3.6.4) yields

$$\begin{cases} d\mathbb{E}[X(t)] = (A + \bar{A})\mathbb{E}[X(t)]dt, \quad t \geqslant 0, \\ \mathbb{E}[X(0)] = x, \end{cases}$$

from which we obtain

$$\mathbb{E}[X(t)] = e^{(A+\bar{A})t}x, \quad t \geqslant 0.$$

Since A, $\bar{A}$ are constant matrices and

$$|\mathbb{E}[X(t)]|^2 \leqslant \mathbb{E}|X(t)|^2 \to 0 \quad \text{as } t \to \infty,$$

the exponential stability of $A + \bar{A}$ follows.

(ii) Let $P \in \mathbb{S}^n$ be a positive definite matrix. By integration by parts,

$$\begin{aligned} \mathbb{E}\langle P\{X(t) &- \mathbb{E}[X(t)]\}, X(t) - \mathbb{E}[X(t)]\rangle \\ &= \mathbb{E}\int_0^t \Big[\langle(PA + A^\top P + C^\top PC)\{X(s) - \mathbb{E}[X(s)]\}, X(s) - \mathbb{E}[X(s)]\rangle \\ &\quad + \langle(C + \bar{C})^\top P(C + \bar{C})\mathbb{E}[X(s)], \mathbb{E}[X(s)]\rangle\Big]ds. \end{aligned} \tag{3.6.8}$$

If (3.6.7) holds, (3.6.8) implies that for some constant $L > 0$,

$$\text{var}\,[X(t)] \leqslant L\int_0^t \text{var}\,[X(s)]ds, \quad \forall t \geqslant 0.$$

Then, by Gronwall's inequality, $\text{var}\,[X(t)] = 0$ and hence

$$X(t) = \mathbb{E}[X(t)] = e^{(A+\bar{A})t}x, \quad t \geqslant 0.$$

So the system $[A, \bar{A}, C, \bar{C}]$ is L^2-exponentially stable since $A + \bar{A}$ is exponentially stable.

If $[A, C]$ is L^2-globally integrable, then by Theorem 1.2.3 of Chap. 1 (with $B = D = 0$), the equation

$$PA + A^\top P + C^\top PC + I = 0 \tag{3.6.9}$$

admits a positive definite solution $P \in \mathbb{S}^n$. With $\rho > 0$ being the smallest eigenvalue of P and $\mu \geqslant 0$ being the largest eigenvalue of $(C + \bar{C})^\top P(C + \bar{C})$, (3.6.8) implies

$$\rho\,\text{var}\,[X(t)] \leqslant -\int_0^t \text{var}\,[X(s)]ds + \mu\int_0^t |\mathbb{E}[X(s)]|^2 ds, \quad t \geqslant 0.$$

By Gronwall's inequality,

$$\text{var}\,[X(t)] \leqslant \frac{\mu}{\rho} \int_0^t e^{\frac{s-t}{\rho}} \big|\mathbb{E}[X(s)]\big|^2 ds, \quad t \geqslant 0. \tag{3.6.10}$$

Since $A + \bar{A}$ is exponentially stable, there exist constants $L, \lambda > 0$ such that

$$\big|\mathbb{E}[X(t)]\big| = |e^{(A+\bar{A})t}x| \leqslant Le^{-\lambda t}, \quad t \geqslant 0. \tag{3.6.11}$$

Combining (3.6.10) and (3.6.11), we can obtain

$$\text{var}\,[X(t)] \leqslant \frac{L\mu}{1-\lambda\rho}\Big[e^{-\lambda t} - e^{-\rho^{-1}t}\Big], \quad t \geqslant 0.$$

This results in

$$
\begin{aligned}
\mathbb{E}|X(t)|^2 &= \text{var}\,[X(t)] + \big|\mathbb{E}[X(t)]\big|^2 \\
&\leqslant \frac{L\mu}{1-\lambda\rho}\Big[e^{-\lambda t} - e^{-\rho^{-1}t}\Big] + L^2 e^{-2\lambda t}, \quad t \geqslant 0,
\end{aligned}
$$

from which the L^2-exponential stability of $[A, \bar{A}, C, \bar{C}]$ follows. $\qquad\square$

It is worth noting that in high dimensions, the L^2-exponential stability, the L^2-global integrability, and the L^2-asymptotic stability of system $[A, \bar{A}, C, \bar{C}]$ are not equivalent. However, in the one-dimensional case, we have the following equivalence result.

Proposition 3.6.6 *For the one-dimensional system*

$$
\begin{cases}
dX(t) = \{aX(t) + \bar{a}\mathbb{E}[X(t)]\}dt + \{cX(t) + \bar{c}\mathbb{E}[X(t)]\}dW, \quad t \geqslant 0, \\
X(0) = x,
\end{cases}
$$

where $a, \bar{a}, c, \bar{c}$ are real numbers, the following statements are equivalent:

(i) *It is L^2-exponentially stable.*
(ii) *It is L^2-globally integrable.*
(iii) *It is L^2-asymptotically stable.*
(iv) *$a + \bar{a} < 0$, and either $2a + c^2 < 0$ or $c + \bar{c} = 0$.*

Proof It suffices to prove the implication (iii) $\Rightarrow$ (iv). By Itô's rule, we have

$$\mathbb{E}|X(t)|^2 = x^2 + \int_0^t \Big\{(2a + c^2)\mathbb{E}|X(s)|^2 + (2\bar{a} + \bar{c}^2 + 2c\bar{c})|\mathbb{E}[X(s)]|^2\Big\}ds.$$

So $\mathbb{E}|X(t)|^2$ satisfies the following ODE:

$$\begin{cases} \dot{y}(t) = (2a + c^2)y(t) + (2\bar{a} + \bar{c}^2 + 2c\bar{c})|\mathbb{E}[X(t)]|^2, & t \geqslant 0, \\ y(0) = x^2. \end{cases}$$

Thus, noting that $\mathbb{E}[X(t)] = e^{(a+\bar{a})t}x$, we have by the variation of constants formula that

$$\mathbb{E}|X(t)|^2 = x^2 e^{(2a+c^2)t} + x^2 \big[2\bar{a} - c^2 + (c + \bar{c})^2 \big]e^{(2a+c^2)t} \int_0^t e^{(2\bar{a}-c^2)s}\,ds$$

$$= x^2 e^{(2a+c^2)t} + x^2 (2\bar{a} - c^2)e^{(2a+c^2)t} \int_0^t e^{(2\bar{a}-c^2)s}\,ds$$

$$+ x^2 (c + \bar{c})^2 e^{(2a+c^2)t} \int_0^t e^{(2\bar{a}-c^2)s}\,ds$$

$$= x^2 e^{2(a+\bar{a})t} + x^2 (c + \bar{c})^2 e^{(2a+c^2)t} \int_0^t e^{(2\bar{a}-c^2)s}\,ds. \tag{3.6.12}$$

Since the system $[a, \bar{a}, c, \bar{c}]$ is L^2-asymptotically stable, Theorem 3.6.5(i) shows that $a + \bar{a} < 0$. Then, from (3.6.12) we further have

$$(c + \bar{c})^2 e^{(2a+c^2)t} \int_0^t e^{(2\bar{a}-c^2)s}\,ds \to 0 \quad \text{as } t \to \infty. \tag{3.6.13}$$

It is not hard to see that (3.6.13) holds only if $2a + c^2 < 0$ or $c + \bar{c} = 0$.   □

For the $L^2_{Q,\bar{Q}}$-global integrability of system $[A, \bar{A}, C, \bar{C}]$, we have the following result.

Theorem 3.6.7 *Let (A1)′–(A2)′ hold. If $[A, \bar{A}, C, \bar{C}]$ is $L^2_{Q,\bar{Q}}$-globally integrable, then $A + \bar{A}$ is $L^2_{Q+\bar{Q}}$-globally integrable, i.e.,*

$$\int_0^\infty \big|(Q + \bar{Q})^{\frac{1}{2}}e^{(A+\bar{A})t}\big|^2\,dt < \infty. \tag{3.6.14}$$

Conversely, if (3.6.14) holds, then $[A, \bar{A}, C, \bar{C}]$ is $L^2_{Q,\bar{Q}}$-globally integrable, provided either (3.6.7) holds, or $[A, C]$ is L^2_Q-globally integrable and

$$\mathcal{N}(Q + \bar{Q}) \subseteq \mathcal{N}(C + \bar{C}), \tag{3.6.15}$$

where $\mathcal{N}(G)$ stands for the null space of a matrix G.

Proof Suppose that (3.6.3) holds. Since $Q \geqslant 0$, we have for any $x \in \mathbb{R}^n$,

$$\int_0^\infty \left| (Q + \bar{Q})^{\frac{1}{2}} e^{(A+\bar{A})t} x \right|^2 dt$$

$$= \int_0^\infty \langle (Q + \bar{Q})\mathbb{E}[X(t)], \mathbb{E}[X(t)] \rangle dt$$

$$\leqslant \int_0^\infty \left(\mathbb{E}\langle QX(t), X(t) \rangle + \langle \bar{Q}\mathbb{E}[X(t)], \mathbb{E}[X(t)] \rangle \right) dt$$

$$= \mathbb{E} \int_0^\infty \left(|Q^{\frac{1}{2}}\{X(t) - \mathbb{E}[X(t)]\}|^2 + |(Q + \bar{Q})^{\frac{1}{2}}\mathbb{E}[X(t)]|^2 \right) dt < \infty,$$

and (3.6.14) follows.

Conversely, let us suppose that (3.6.14) holds. Recall that (3.6.8) is true for any $P \in \mathbb{S}^n$. So if (3.6.7) holds, the same argument in the proof of Theorem 3.6.5(ii) shows that

$$X(t) = \mathbb{E}[X(t)] = e^{(A+\bar{A})t} x, \quad t \geqslant 0.$$

Consequently,

$$\mathbb{E} \int_0^\infty \left(|Q^{\frac{1}{2}}\{X(t) - \mathbb{E}[X(t)]\}|^2 + |(Q + \bar{Q})^{\frac{1}{2}}\mathbb{E}[X(t)]|^2 \right) dt$$

$$= \int_0^\infty |(Q + \bar{Q})^{\frac{1}{2}} e^{(A+\bar{A})t} x|^2 dt < \infty.$$

If (3.6.15) holds and $[A, C]$ is L_Q^2-globally integrable, then proceeding similarly to the proof of Theorem 3.2.3 in [48, Chap. 3] (with Λ replaced by Q), one can show that there exists a matrix $P \geqslant 0$ such that

$$PA + A^\top P + C^\top PC + Q = 0.$$

With this P, (3.6.8) yields

$$0 \leqslant \mathbb{E}\langle P\{X(t) - \mathbb{E}[X(t)]\}, X(t) - \mathbb{E}[X(t)] \rangle$$

$$= \mathbb{E} \int_0^t \Big[-\langle Q\{X(s) - \mathbb{E}[X(s)]\}, X(s) - \mathbb{E}[X(s)] \rangle$$

$$+ \langle (C + \bar{C})^\top P(C + \bar{C})\mathbb{E}[X(s)], \mathbb{E}[X(s)] \rangle \Big] ds.$$

Now, the condition (3.6.15) implies that

$$\langle (C + \bar{C})^\top P(C + \bar{C})y, y \rangle \leqslant L\langle (Q + \bar{Q})y, y \rangle, \quad \forall y \in \mathbb{R}^n,$$

for some $L > 0$. Thus,

$$\mathbb{E} \int_0^t \langle Q\{X(s) - \mathbb{E}[X(s)]\}, X(s) - \mathbb{E}[X(s)]\rangle ds$$

$$\leqslant L \int_0^t \langle (Q + \bar{Q})\mathbb{E}[X(s)], \mathbb{E}[X(s)]\rangle ds, \quad \forall t \geqslant 0.$$

Consequently,

$$\mathbb{E} \int_0^\infty \left(|Q^{\frac{1}{2}}\{X(s) - \mathbb{E}[X(s)]\}|^2 + |(Q + \bar{Q})^{\frac{1}{2}}\mathbb{E}[X(s)]|^2 \right) ds$$

$$= \mathbb{E} \int_0^\infty \left(\langle Q\{X(s) - \mathbb{E}[X(s)]\}, X(s) - \mathbb{E}[X(s)] \rangle \right.$$

$$\left. + \langle (Q + \bar{Q})\mathbb{E}[X(s)], \mathbb{E}[X(s)]\rangle \right) ds$$

$$\leqslant (L + 1) \int_0^\infty \langle (Q + \bar{Q})\mathbb{E}[X(s)], \mathbb{E}[X(s)]\rangle ds$$

$$= (L + 1) \int_0^\infty |(Q + \bar{Q})^{\frac{1}{2}} e^{(A+\bar{A})s} x|^2 ds < \infty.$$

This means that the system is $L^2_{Q,\bar{Q}}$-globally integrable.   □

We now return to the controlled linear mean-field SDE (3.6.1), which we denote by $[A, \bar{A}, C, \bar{C}; B, \bar{B}, D, \bar{D}]$ for simplicity of notation.

Definition 3.6.8 Let $(A1)'$–$(A2)'$ hold. The system $[A, \bar{A}, C, \bar{C}; B, \bar{B}, D, \bar{D}]$ is said to be *MF-$L^2_{Q,\bar{Q}}$-stabilizable* if there exist matrices $K, \bar{K} \in \mathbb{R}^{m \times n}$ such that for any $x \in \mathbb{R}^n$, if $X^{K,\bar{K}}$ is the solution of

$$\begin{cases} dX(t) = \{(A + BK)X(t) + [\bar{A} + \bar{B}\bar{K} + B(\bar{K} - K)]\mathbb{E}[X(t)]\}dt \\ \qquad\quad + \{(C + DK)X(t) + [\bar{C} + \bar{D}\bar{K} + D(\bar{K} - K)]\mathbb{E}[X(t)]\}dW, \\ X(0) = x, \end{cases}$$

and

$$u^{K,\bar{K}} = K\{X^{K,\bar{K}} - \mathbb{E}[X^{K,\bar{K}}]\} + \bar{K}\mathbb{E}[X^{K,\bar{K}}],$$

then

$$\mathbb{E} \int_0^\infty \left\{ \langle QX^{K,\bar{K}}(t), X^{K,\bar{K}}(t)\rangle + \langle \bar{Q}\mathbb{E}[X^{K,\bar{K}}(t)], \mathbb{E}[X^{K,\bar{K}}(t)]\rangle \right.$$

$$\left. + |u^{K,\bar{K}}(t)|^2 \right\} dt < \infty.$$

In this case, the pair $(K, \bar{K})$ is called an *MF-$L^2_{Q,\bar{Q}}$-stabilizer* of the system. When the stronger condition

$$\mathbb{E} \int_0^\infty \left[|X^{K,\bar{K}}(t)|^2 + |u^{K,\bar{K}}(t)|^2 \right] dt < \infty$$

holds, we say that the system is *MF-L^2-stabilizable* and call $(K, \bar{K})$ an *MF-L^2-stabilizer* of the system.

Definition 3.6.9 Let $(A1)'$–$(A2)'$ hold. The system $[A, \bar{A}, C, \bar{C}; B, \bar{B}, D, \bar{D}]$ is said to be $L^2_{Q,\bar{Q}}$-*stabilizable* if there exists a matrix $K \in \mathbb{R}^{m \times n}$ such that for any $x \in \mathbb{R}^n$, if X^K is the solution of

$$\begin{cases} dX(t) = \{(A + BK)X(t) + (\bar{A} + \bar{B}K)\mathbb{E}[X(t)]\}dt \\ \qquad\quad + \{(C + DK)X(t) + (\bar{C} + \bar{D}K)\mathbb{E}[X(t)]\}dW, \\ X(0) = x, \end{cases}$$

and

$$u^K(t) = KX^K(t), \quad t \geqslant 0,$$

then

$$\mathbb{E} \int_0^\infty \left\{ \langle QX^K(t), X^K(t) \rangle + \langle \bar{Q}\mathbb{E}[X^K(t)], \mathbb{E}[X^K(t)] \rangle + |u^K(t)|^2 \right\} dt < \infty.$$

In this case, K is called an $L^2_{Q,\bar{Q}}$-*stabilizer* of the system. In the special case of $\bar{Q} = 0$, we simply say that the system is L^2_Q-*stabilizable* and call K an L^2_Q-*stabilizer*. When the stronger condition

$$\mathbb{E} \int_0^\infty \left[|X^K(t)|^2 + |u^K(t)|^2 \right] dt < \infty$$

holds, we say that the system is L^2-*stabilizable* and call K an L^2-*stabilizer* of the system.

The importance of the notions defined above is that when the system $[A, \bar{A}, C, \bar{C};$ $B, \bar{B}, D, \bar{D}]$ is MF-$L^2_{Q,\bar{Q}}$-stabilizable, the admissible control set $\mathcal{U}_{ad}(x)$ is nonempty for every initial state $x \in \mathbb{R}^n$, since $u^{K,\bar{K}} \in \mathcal{U}_{ad}(x)$. In particular, $\mathcal{U}_{ad}(x)$ is nonempty for all $x \in \mathbb{R}^n$ if the system is MF-L^2-stabilizable.

It is easily seen that if $[A, \bar{A}, C, \bar{C}; B, \bar{B}, D, \bar{D}]$ is MF-$L^2_{Q,\bar{Q}}$-stabilizable, then with $(K, \bar{K})$ being an stabilizer, the uncontrolled system

$$[A + BK, \bar{A} + \widehat{B}\bar{K} - BK, C + DK, \bar{C} + \widehat{D}\bar{K} - DK], \tag{3.6.16}$$

where $\widehat{B} = B + \bar{B}$ and $\widehat{D} = D + \bar{D}$, is $L^2_{Q,\bar{Q}}$-globally integrable. Moreover, when the mean-field terms are absent, the notion of L^2-stabilizability defined here coincides with that defined in Chap. 1, Sect. 1.2.

Note that if the system (3.6.1) is L^2-stabilizable, then it is also MF-L^2-stabilizable. In fact, if K is an L^2-stabilizer, then (K, K) is an MF-L^2-stabilizer. The following example shows that in general, the MF-L^2-stabilizability does not imply the L^2-stabilizability.

Example 3.6.10 Consider the one-dimensional controlled mean-filed SDE

$$
\begin{cases}
dX(t) = \{aX(t) + \bar{a}\mathbb{E}[X(t)] + bu(t) + \bar{b}\mathbb{E}[u(t)]\}dt \\
\qquad + \{cX(t) + \bar{c}\mathbb{E}[X(t)] + du(t) + \bar{d}\mathbb{E}[u(t)]\}dW, \qquad (3.6.17) \\
X(0) = x,
\end{cases}
$$

where the coefficients

$$a = -1, \quad \bar{a} = 2, \quad b = 1, \quad \bar{b} = 0, \quad c = 1, \quad \bar{c} = 0, \quad d = -1, \quad \bar{d} = 1.$$

To see that the system (3.6.17) is MF-L^2-stabilizable, we need only find $k, \bar{k} \in \mathbb{R}$ such that the closed-loop system

$$
\begin{aligned}
dX(t) = &\{(a + bk)X(t) + [\bar{a} + \bar{b}\bar{k} + b(\bar{k} - k)]\mathbb{E}[X(t)]\}dt \\
&+ \{(c + dk)X(t) + [\bar{c} + \bar{d}\bar{k} + d(\bar{k} - k)]\mathbb{E}[X(t)]\}dW
\end{aligned}
$$

is L^2-globally integrable. By Proposition 3.6.6, the above closed-loop system is L^2-globally integrable if and only if

$$
\begin{aligned}
a + \bar{a} + (b + \bar{b})\bar{k} &= \bar{k} + 1 < 0, \\
2(a + bk) + (c + dk)^2 &= k^2 - 1 < 0.
\end{aligned}
$$

So $(k, \bar{k}) = (0, -2)$ is an MF-L^2-stabilizer.

To see that the system (3.6.17) is not L^2-stabilizable, we need verify that for any $k \in \mathbb{R}$, the system

$$
\begin{aligned}
dX(t) = &\{(a + bk)X(t) + (\bar{a} + \bar{b}k)\mathbb{E}[X(t)]\}dt \\
&+ \{(c + dk)X(t) + (\bar{c} + \bar{d}k)\mathbb{E}[X(t)]\}dW
\end{aligned}
$$

is not L^2-globally integrable. This is equivalent to verifying that the system of inequalities

$$(a + bk) + (\bar{a} + \bar{b}k) = k + 1 < 0,$$
$$2(a + bk) + (c + dk)^2 = k^2 - 1 < 0,$$

has no solutions, which is trivial.

Now we present a result concerning the MF-$L^2_{Q,\bar{Q}}$-stabilizability of (3.6.1). We shall use the following notation:

$$\widehat{A} = A + \bar{A}, \quad \widehat{B} = B + \bar{B}, \quad \widehat{C} = C + \bar{C}, \quad \widehat{D} = D + \bar{D}, \quad \widehat{Q} = Q + \bar{Q}.$$

Theorem 3.6.11 *Let (A1)′–(A2)′ hold.*

(i) *If the system (3.6.1) is MF-$L^2_{Q,\bar{Q}}$-stabilizable, then for some $\bar{K} \in \mathbb{R}^{m \times n}$,*

$$\int_0^\infty \left| \widehat{Q}^{\frac{1}{2}} e^{(\widehat{A} + \widehat{B}\bar{K})t} \right|^2 dt < \infty. \tag{3.6.18}$$

(ii) *Suppose that there exists $\bar{K} \in \mathbb{R}^{m \times n}$ such that (3.6.18) and*

$$\mathcal{N}(\widehat{Q}) \subseteq \mathcal{N}(\widehat{C} + \widehat{D}\bar{K}) \tag{3.6.19}$$

hold. If the controlled SDE system

$$dX(s) = [AX(s) + Bu(s)]ds + [CX(s) + Du(s)]dW \tag{3.6.20}$$

is L^2_Q-stabilizable, then the system (3.6.1) is MF-$L^2_{Q,\bar{Q}}$-stabilizable.

(iii) *Suppose that there exists $\bar{K} \in \mathbb{R}^{m \times n}$ such that (3.6.18) and*

$$\widehat{C} + \widehat{D}\bar{K} = 0 \tag{3.6.21}$$

hold. Then the system (3.6.1) is MF-$L^2_{Q,\bar{Q}}$-stabilizable.

Proof The proof of Theorem 3.6.11 follows easily from Theorem 3.6.7. The details are left to the reader. □

Corollary 3.6.12 *Let (A1)′–(A2)′ hold.*

(i) *If the system (3.6.1) is MF-L^2-stabilizable, then there exists a matrix $\bar{K} \in \mathbb{R}^{m \times n}$ such that*

$$\sigma(\widehat{A} + \widehat{B}\bar{K}) \subseteq \mathbb{C}^- \equiv \{z = \alpha + i\beta \in \mathbb{C} \mid \alpha < 0\}, \tag{3.6.22}$$

where $\sigma(\widehat{A} + \widehat{B}\bar{K})$ is the spectrum of $\widehat{A} + \widehat{B}\bar{K}$.

(ii) *Suppose that (3.6.22) holds for some $\bar{K} \in \mathbb{R}^{m \times n}$, and that the system (3.6.20) is L^2-stabilizable. Then the controlled mean-field SDE system (3.6.1) is MF-L^2-stabilizable.*

(iii) *Suppose that there exists a matrix* $\bar{K} \in \mathbb{R}^{m \times n}$ *such that (3.6.21) and (3.6.22)*
    hold. Then the controlled mean-field SDE system (3.6.1) is MF-L^2-stabilizable.

Note that the assumptions in Corollary 3.6.12(ii) do not involve $\bar{C}$ or $\bar{D}$. However, the condition (3.6.21) in Corollary 3.6.12(iii) involves both $\bar{C}$ and $\bar{D}$. We point out that (3.6.21) means that

$$\mathscr{R}(\widehat{C}) \subseteq \mathscr{R}(\widehat{D}). \tag{3.6.23}$$

In the case of $m < n$, the above could be a big restriction on $\widehat{C}$ and $\widehat{D}$.

In Corollary 3.6.12(iii), the matrix $\bar{K} \in \mathbb{R}^{m \times n}$ should satisfy (3.6.21) and (3.6.22) simultaneously. The solution of (3.6.21) is given by

$$\bar{K} = -\widehat{D}^\dagger \widehat{C} + (I - \widehat{D}^\dagger \widehat{D})\widetilde{K},$$

where $\widetilde{K} \in \mathbb{R}^{m \times n}$ is arbitrary. Thus, in order for (3.6.22) to hold, we need

$$\sigma\Big(\widehat{A} + \widehat{B}[-\widehat{D}^\dagger \widehat{C} + (I - \widehat{D}^\dagger \widehat{D})\widetilde{K}]\Big) \subseteq \mathbb{C}^-,$$

which means that the ODE system

$$\dot{X}(t) = (\widehat{A} - \widehat{B}\widehat{D}^\dagger \widehat{C})X(t) + \widehat{B}(I - \widehat{D}^\dagger \widehat{D})u(t), \quad t \geqslant 0 \tag{3.6.24}$$

is stabilizable. Hence, we obtain the following result.

Proposition 3.6.13 *Let (A1)′–(A2)′ and (3.6.23) hold. Then the system (3.6.1) is MF-L^2-stabilizable if the ODE system (3.6.24) is stabilizable, which is the case, if, in particular, $m = n$, $\widehat{D}$ is invertible, and*

$$\sigma(\widehat{A} - \widehat{B}\widehat{D}^{-1}\widehat{C}) \subseteq \mathbb{C}^-. \tag{3.6.25}$$

From (3.6.25) it seems that the MF-L^2-stabilizability of the mean-filed SDE system (3.6.1) could have nothing to do with the stabilizability of the SDE system (3.6.20). In the case that $\bar{A} = \bar{C} = 0$ and $\bar{B} = \bar{D} = 0$, we have the following controlled linear SDE:

$$dX(t) = [AX(t) + Bu(t)]dt + [CX(t) + Du(t)]dW, \quad t \geqslant 0.$$

Suppose that $m = n$ and D^{-1} exists. Then the condition (3.6.25) becomes

$$\sigma(A - BD^{-1}C) \subseteq \mathbb{C}^-. \tag{3.6.26}$$

In this case, if we take
$$u(t) = -D^{-1}CX(t), \quad t \geqslant 0,$$

then the closed-loop system becomes

$$dX(t) = (A - BD^{-1}C)X(t)dt, \quad t \geqslant 0,$$

which is exponentially stable if (3.6.26) holds.

3.6.2 Stochastic LQ Problems

We now return to Problem (MFLQ)$_\infty$. In order for the admissible control set $\mathcal{U}_{ad}(x)$ to be nonempty for each initial state $x \in \mathbb{R}^n$, we make the following assumption.

(A3)′ The controlled ODE system

$$\dot{X}(t) = \widehat{A}X(t) + \widehat{B}u(t), \quad t \geqslant 0$$

is stabilizable, and the controlled SDE system

$$dX(t) = [AX(t) + Bu(t)]dt + [CX(t) + Du(t)]dW, \quad t \geqslant 0$$

is L^2-stabilizable.

From Corollary 3.6.12 (ii), we know that under the assumptions (A1)′–(A3)′, the system (3.6.1) is MF-L^2-stabilizable. It is possible to relax (A3)′ in various ways. However, for simplicity of presentation, we would like to keep (A3)′ in the sequel.

Let us first introduce the following notation (similar to that used for the finite-horizon problem):

$$\widehat{A} = A + \bar{A}, \quad \widehat{B} = B + \bar{B}, \quad \widehat{C} = C + \bar{C}, \quad \widehat{D} = D + \bar{D},$$
$$\widehat{Q} = Q + \bar{Q}, \quad \widehat{R} = R + \bar{R}, \quad \widehat{G} = G + \bar{G},$$

and for constant matrices $P, \Pi \in \mathbb{S}^n$,

$$\mathcal{Q}(P) = PA + A^\top P + C^\top PC + Q,$$
$$\mathcal{S}(P) = B^\top P + D^\top PC,$$
$$\mathcal{R}(P) = R + D^\top PD,$$

and

$$\widehat{\mathcal{Q}}(P, \Pi) = \Pi\widehat{A} + \widehat{A}^\top \Pi + \widehat{C}^\top P\widehat{C} + \widehat{Q},$$
$$\widehat{\mathcal{S}}(P, \Pi) = \widehat{B}^\top \Pi + \widehat{D}^\top P\widehat{C},$$
$$\widehat{\mathcal{R}}(P) = \widehat{R} + \widehat{D}^\top P\widehat{D}.$$

Note that different from (3.4.3) and (3.4.4), all the matrices in the above are constant-valued and for simplicity we take $S = \bar{S} = 0$.

Theorem 3.6.14 *Let (A1)'–(A3)' hold. We have the following results:*

(i) *For every initial state $x \in \mathbb{R}^n$, Problem (MFLQ)$_\infty$ admits a unique optimal control $u^* \in \mathcal{U}_{ad}(x)$.*

(ii) *The algebraic Riccati equations (AREs, for short)*

$$\mathcal{Q}(P) - \mathcal{S}(P)^\top \mathcal{R}(P)^{-1} \mathcal{S}(P) = 0, \qquad (3.6.27)$$

$$\widehat{\mathcal{Q}}(P, \Pi) - \widehat{\mathcal{S}}(P, \Pi)^\top \widehat{\mathcal{R}}(P)^{-1} \widehat{\mathcal{S}}(P, \Pi) = 0, \qquad (3.6.28)$$

admit a solution pair $(P, \Pi) \in \bar{\mathbb{S}}^n_+ \times \bar{\mathbb{S}}^n_+$.

(iii) *The pair $(\Theta, \widehat{\Theta})$ defined by*

$$\Theta \triangleq -\mathcal{R}(P)^{-1} \mathcal{S}(P), \quad \widehat{\Theta} \triangleq -\widehat{\mathcal{R}}(P)^{-1} \widehat{\mathcal{S}}(P, \Pi)$$

is an MF-$L^2_{Q,\bar{Q}}$-stabilizer of the system (3.6.1).

(iv) *The optimal control u^* for the initial state x admits the following state feedback representation:*

$$u^*(t) = \Theta\{X(t) - \mathbb{E}[X(t)]\} + \widehat{\Theta}\mathbb{E}[X(t)], \quad t \geqslant 0,$$

where X is the solution to the following mean-field closed-loop system:

$$\begin{cases} dX(t) = \{(A + B\Theta)X(t) + (\bar{A} + \widehat{B}\widehat{\Theta} - B\Theta)\mathbb{E}[X(t)]\}dt \\ \qquad\quad + \{(C + D\Theta)X(t) + (\bar{C} + \widehat{D}\widehat{\Theta} - D\Theta)\mathbb{E}[X(t)]\}dW, \quad (3.6.29) \\ X(0) = x. \end{cases}$$

(v) *The value function of Problem (MFLQ)$_\infty$ is given by*

$$V(x) \triangleq \inf_{u \in \mathcal{U}_{ad}(x)} J(x; u) = \langle \Pi x, x \rangle, \quad \forall x \in \mathbb{R}^n.$$

Proof Under (A1)'–(A3)', the admissible control set $\mathcal{U}_{ad}(x)$ is nonempty and convex for each $x \in \mathbb{R}^n$. Writing the cost functional $J(x; u)$ as

$$J(x; u) = \mathbb{E} \int_0^\infty \Big\{ \langle Q\{X(t) - \mathbb{E}[X(t)]\}, X(t) - \mathbb{E}[X(t)] \rangle + \langle \widehat{Q}\mathbb{E}[X(t)], \mathbb{E}[X(t)] \rangle $$
$$+ \langle R\{u(t) - \mathbb{E}[u(t)]\}, u(t) - \mathbb{E}[u(t)] \rangle + \langle \widehat{R}\mathbb{E}[u(t)], \mathbb{E}[u(t)] \rangle \Big\} dt,$$

we see, by the assumption (A2)', that

$$J(x; u) \geqslant \delta \mathbb{E} \int_0^\infty |u(t)|^2 dt, \quad \forall u \in \mathcal{U}_{ad}(x)$$

for some $\delta > 0$. Thus, the map $u \mapsto J(x; u)$ is strictly convex on $\mathcal{U}_{ad}(x)$, and hence the optimal control of Problem $(MFLQ)_\infty$ for x is unique.

To prove the existence of an optimal control as well as the assertions (ii)–(v), define for each $T > 0$ the cost functional

$$J_T(x; u) = \mathbb{E} \int_0^T \Big\{ \langle QX(t), X(t) \rangle + \langle \bar{Q}\mathbb{E}[X(t)], \mathbb{E}[X(t)] \rangle$$
$$+ \langle Ru(t), u(t) \rangle + \langle \bar{R}\mathbb{E}[u(t)], \mathbb{E}[u(t)] \rangle \Big\} dt$$

and consider the following problem.

Problem $(MFLQ)_T$. For given $x \in \mathbb{R}^n$, find a control $u_T^* \in \mathcal{U}[0, T]$ such that

$$J_T(x; u_T^*) = \inf_{u \in \mathcal{U}[0,T]} J_T(x; u) \equiv V_T(x).$$

By the results of Sect. 3.4 in this chapter, Problem $(MFLQ)_T$ admits a unique optimal control u_T^* and

$$V_T(x) = J_T(x; u_T^*) = \langle \Pi_T(0)x, x \rangle, \quad \forall x \in \mathbb{R}^n,$$

where $\Pi_T(t) \geqslant 0$ is the solution to the Riccati equation

$$\begin{cases} \dot{\Pi}_T(t) + \mathcal{Q}(P_T(t), \Pi_T(t)) - \widehat{\mathcal{S}}(P_T(t), \Pi_T(t))^\top \\ \quad \times \widehat{\mathcal{R}}(P_T(t))^{-1}\widehat{\mathcal{S}}(P_T(t), \Pi_T(t)) = 0, \quad t \in [0, T], \\ \Pi_T(T) = 0, \end{cases}$$

with $P_T(t) \geqslant 0$ being the solution to the Riccati equation

$$\begin{cases} \dot{P}_T(t) + \mathcal{Q}(P_T(t)) - \mathcal{S}(P_T(t))^\top \mathcal{R}(P_T(t))^{-1}\mathcal{S}(P_T(t)) = 0, \quad t \in [0, T], \\ P_T(T) = 0. \end{cases}$$

Furthermore, with the notation

$$\Theta_T(t) = -\mathcal{R}(P_T(t))^{-1}\mathcal{S}(P_T(t)),$$
$$\widehat{\Theta}_T(t) = -\widehat{\mathcal{R}}(P_T(t))^{-1}\widehat{\mathcal{S}}(P_T(t), \Pi_T(t)),$$

the optimal control u_T^* admits the following state feedback representation:

$$u_T^*(t) = \Theta_T(t)\{X_T(t) - \mathbb{E}[X_T(t)]\} + \widehat{\Theta}_T(t)\mathbb{E}[X_T(t)], \quad t \in [0, T],$$

where X_T is the solution to the following closed-loop system over $[0, T]$:

$$\begin{cases} dX_T(t) = \left\{ \left[A + B\Theta_T(t) \right] X_T(t) + \left[\bar{A} + \widehat{B}\widehat{\Theta}_T(t) - B\Theta_T(t) \right] \mathbb{E}[X_T(t)] \right\} dt \\ \qquad + \left\{ \left[C + D\Theta_T(t) \right] X_T(t) + \left[\bar{C} + \widehat{D}\widehat{\Theta}_T(t) - D\Theta_T(t) \right] \mathbb{E}[X_T(t)] \right\} dW, \\ X_T(0) = x. \end{cases}$$

To see that the ARE (3.6.27) admits a solution $P \in \bar{\mathbb{S}}_+^n$, let us consider the state equation

$$\begin{cases} dX(t) = [AX(t) + Bu(t)]dt + [CX(t) + Du(t)]dW, \quad t \geqslant 0, \\ X(0) = x, \end{cases}$$

and the following cost functionals:

$$\widehat{J}_T(x; u) = \mathbb{E} \int_0^T \left[\langle QX(t), X(t) \rangle + \langle Ru(t), u(t) \rangle \right] dt,$$
$$\widehat{J}(x; u) = \mathbb{E} \int_0^\infty \left[\langle QX(t), X(t) \rangle + \langle Ru(t), u(t) \rangle \right] dt.$$

By Theorem 1.1.15 of Chap. 1,

$$\langle P_T(0)x, x \rangle = \inf_{u \in \mathcal{U}[0,T]} \widehat{J}_T(x; u) \equiv \widehat{V}_T(x), \quad \forall x \in \mathbb{R}^n.$$

Since the controlled SDE system

$$dX(t) = [AX(t) + Bu(t)]dt + [CX(t) + Du(t)]dW, \quad t \geqslant 0$$

is L^2-stabilizable, we have

$$\widehat{V}(x) \triangleq \inf_{u \in L_{\mathbb{F}}^2(\mathbb{R}^m)} \widehat{J}(x; u) < \infty.$$

It is not hard to see that for any $0 < T < T'$,

$$\widehat{V}_T(x) \leqslant \widehat{V}_{T'}(x) \leqslant \widehat{V}(x), \quad \forall x \in \mathbb{R}^n,$$

from which we conclude that the limit

$$P \triangleq \lim_{T \to \infty} P_T(0) \geqslant 0$$

exists and is finite. By Lemma 3.3.4 of [48, Chap. 3], P is a solution of (3.6.27). Further, proceeding exactly as in the proof of Lemma 3.3.4 in [48, Chap. 3], we can show that

$$P_T(T - s) = P_{T'}(T' - s), \quad \forall s \in [0, T],$$

from which it follows that $P_T(0) = P_{T+t}(t)$ for all $t \geqslant 0$, and hence

$$\lim_{T \to \infty} P_T(t) = P, \quad \forall t \geqslant 0. \tag{3.6.30}$$

By a similar argument, we can show that

$$\Pi \triangleq \lim_{T \to \infty} \Pi_T(0) \geqslant 0$$

exists and is a solution of (3.6.28), and that

$$\lim_{T \to \infty} \Pi_T(t) = \Pi, \quad \forall t \geqslant 0. \tag{3.6.31}$$

From (3.6.30) and (3.6.31) we obtain

$$\lim_{T \to \infty} \Theta_T(t) = - \lim_{T \to \infty} \mathcal{R}(P_T(t))^{-1} \mathcal{S}(P_T(t)) = -\mathcal{R}(P)^{-1} \mathcal{S}(P) = \Theta,$$
$$\lim_{T \to \infty} \widehat{\Theta}_T(t) = - \lim_{T \to \infty} \widehat{\mathcal{R}}(P_T(t))^{-1} \widehat{\mathcal{S}}(P_T(t), \Pi_T(t)) = \widehat{\Theta}.$$

Thus, for any $t \geqslant 0$, we have almost surely

$$\lim_{T \to \infty} X_T(t) = X(t),$$
$$\lim_{T \to \infty} u_T^*(t) = \Theta \{ X(t) - \mathbb{E}[X(t)] \} + \widehat{\Theta} \mathbb{E}[X(t)] \equiv u^*(t),$$

where X is the solution of (3.6.29). Recall that

$$\langle \Pi_T(0)x, x \rangle = J_T(x; u_T^*)$$
$$= \mathbb{E} \int_0^T \Big\{ \langle Q X_T(t), X_T(t) \rangle + \langle \bar{Q} \mathbb{E}[X_T(t)], \mathbb{E}[X_T(t)] \rangle$$
$$+ \langle R u_T^*(t), u_T^*(t) \rangle + \langle \bar{R} \mathbb{E}[u_T^*(t)], \mathbb{E}[u_T^*(t)] \rangle \Big\} dt.$$

Letting $T \to \infty$, we have by Fatou's Lemma,

$$V(x) \geqslant \lim_{T \to \infty} V_T(x) = \langle \Pi x, x \rangle$$
$$\geqslant \mathbb{E} \int_0^\infty \Big\{ \langle Q X(t), X(t) \rangle + \langle \bar{Q} \mathbb{E}[X(t)], \mathbb{E}[X(t)] \rangle$$
$$+ \langle R u^*(t), u^*(t) \rangle + \langle \bar{R} \mathbb{E}[u^*(t)], \mathbb{E}[u^*(t)] \rangle \Big\} dt \geqslant V(x).$$

This shows that u^* is an optimal control for x and that the assertions (ii)–(v) hold. $\qquad \square$

Bibliography

1. Ahmed, N.U.: Nonlinear diffusion governed by McKean-Vlasov equation on Hilbert space and optimal control. SIAM J. Control Optim. **46**, 356–378 (2007)
2. Ahmed, N.U., Ding, X.: Controlled McKean-Vlasov equations. Commun. Appl. Anal. **5**, 183–206 (2001)
3. Ait Rami, M., Moore, J.B., Zhou, X.Y.: Indefinite stochastic linear quadratic control and generalized differential Riccati equation. SIAM J. Control Optim. **40**, 1296–1311 (2001)
4. Andersson, D., Djehiche, B.: A maximum principle for SDEs of mean-field type. Appl. Math. Optim. **63**, 341–356 (2011)
5. Bellman, R., Glicksberg, I., Gross, O.: Some Aspects of the Mathematical Theory of Control Processes. RAND Corporation, Santa Monica, CA (1958)
6. Berkovitz, L.D.: Lectures on differential games. In: Kuhn, H.W., Szego, G.P. (eds.) Differential Games and Related Topics, North-Holland, Amsterdam, The Netherlands, pp. 3–45 (1971)
7. Bernhard, P.: Linear-quadratic, two-person, zero-sum differential games: necessary and sufficient conditions. J. Optim. Theory Appl. **27**, 51–69 (1979)
8. Bensoussan, A., Sung, K.C.J., Yam, S.C.P., Yung, S.P.: Linear-quadratic mean field games. J. Optim. Theory Appl. **169**, 496–529 (2016)
9. Bismut, J.M.: Linear quadratic optimal stochastic control with random coefficients. SIAM J. Control Optim. **14**, 419–444 (1976)
10. Borkar, V.S., Kumar, K.S.: McKean-Vlasov limit in portfolio optimization. Stoch. Anal. Appl. **28**, 884–906 (2010)
11. Buckdahn, R., Cardaliaguet, P., Rainer, C.: Nash equilibirum payoffs for nonzero-sum stochastic differential games. SIAM J. Control Optim. **43**, 624–642 (2004)
12. Buckdahn, R., Djehiche, B., Li, J.: A general maximum principle for SDEs of mean-field type. Appl. Math. Optim. **64**, 197–216 (2011)
13. Carmona, R.: Lectures on BSDEs, stochastic control, and stochastic differential games with financial applications. SIAM Book Ser. Financ. Math. **1** (2016)
14. Chan, T.: Dynamics of the McKean-Vlasov equation. Ann. Probab. **22**, 431–441 (1994)
15. Delfour, M.C.: Linear quadratic differential games: Saddle point and Riccati differential equations. SIAM J. Control Optim. **46**, 750–774 (2007)
16. Delfour, M.C., Sbarba, O.D.: Linear quadratic differential games: closed loop saddle points. SIAM J. Control Optim. **47**, 3138–3166 (2009)
17. El Karoui, N., Hamadène, S.: BSDEs and risk-sensitive control, zero-sum and nonzero-sum game problems of stochastic functional differential equations. Stoch. Proc. Appl. **107**, 145–169 (2003)

© The Author(s), under exclusive license to Springer Nature Switzerland AG 2020

J. Sun and J. Yong, *Stochastic Linear-Quadratic Optimal Control Theory: Differential Games and Mean-Field Problems*, SpringerBriefs in Mathematics, https://doi.org/10.1007/978-3-030-48306-7

18. Friedman, A.: Stochastic differential games. J. Differ. Equ. **11**, 79–108 (1972)
19. Hamadène, S.: Backward-forward SDE's and stochastic differential games. Stoch. Proc. Appl. **77**, 1–15 (1998)
20. Hamadène, S.: Nonzero sum linear-quadratic stochastic differential games and backward-forward equations. Stoch. Anal. Appl. **17**, 117–130 (1999)
21. Hamadène, S., Mu, R.: Existence of Nash equilibrium points for Markovian non-zero-sum stochastic differential games with unbounded coeffcients. Stoch. Int. J. Probab. Stoch. Process **87**, 85–111 (2015)
22. Ho, Y.C., Bryson, A.E., Baron, S.: Differential games and optimal pursuit-evasion strategies. IEEE Trans. Automat. Control **10**, 385–389 (1965)
23. Huang, J., Li, X., Yong, J.: A linear-quadratic optimal control problem for mean-field stochastic differential equations in infinite horizon. Math. Control Relat. Fields **5**, 97–139 (2015)
24. Huang, M., Malhamé, R.P., Caines, P.E.: Large population stochastic dynamic games: closed-loop McKean-Vlasov systems and the Nash certainty equivalence principle. Community Inform. Syst. **6**, 221–251 (2006)
25. Ichikawa, A.: Linear quadratic differential games in a Hilbert space. SIAM J. Control Optim. **14**, 120–136 (1976)
26. Kalman, R.E.: Contributions to the theory of optimal control. Bol. Soc. Mat. Mexicana **5**, 102–119 (1960)
27. Karatzas, I., Shreve, S.E.: Brownian Motion and Stochastic Calculus, 2nd edn. Springer, New York (1991)
28. Li, X., Sun, J., Xiong, J.: Linear quadratic optimal control problems for mean-field backward stochastic differential equations. Appl. Math. Optim. **80**, 223–250 (2019)
29. Li, X., Sun, J., Yong, J.: Mean-field stochastic linear quadratic optimal control problems: closed-loop solvability. Probab. Uncertain. Quant. Risk **1**, 2 (2016). https://doi.org/10.1186/s41546-016-0002-3
30. Lin, Q.: A BSDE approach to Nash equilibrium payoffs for stochastic differential games with nonlinear cost functionals. Stoch. Proc. Appl. **122**, 357–385 (2012)
31. Lukes, D.L., Russell, D.L.: A global theory for linear-quadratic differential games. J. Math. Anal. Appl. **33**, 96–123 (1971)
32. Ma, J., Protter, P., Yong, J.: Solving forward-backward stochastic differential equations explicitly–a four-step scheme. Probab. Theory Relat. Fields **98**, 339–359 (1994)
33. Ma, J., Yong, J.: Forward-Backward Stochastic Differential Equations and Their Applications. Lecture Notes in Mathematics, vol. 1702. Springer, New York (1999)
34. McAsey, M., Mou, L.: Generalized Riccati equations arising in stochastic games. Linear Algebra Appl. **416**, 710–723 (2006)
35. Meyer-Brandis, T., Øksendal, B., Zhou, X.Y.: A mean-field stochastic maximum principle via Malliavin calculus. Stochastics **84**, 643–666 (2012)
36. Mou, L., Yong, J.: Two-person zero-sum linear quadratic stochastic differential games by a Hilbert space method. J. Ind. Manag. Optim. **2**, 95–117 (2006)
37. Nash, J.: Non-cooperative games. Ann. Math. **54**, 286–295 (1951)
38. Penrose, R.: A generalized inverse of matrices. Proc. Camb. Philos. Soc. **52**, 17–19 (1955)
39. Pham, H.: Linear quadratic optimal control of conditional McKean-Vlasov equation with random coefficients and applications. Probab. Uncertain. Quant. Risk **1**, 7 (2016). https://doi.org/10.1186/s41546-016-0008-x
40. Pham, H., Basei, M.: Linear-quadratic McKean-Vlasov stochastic control problems with random coefficients on finite and infinite dorizon, and applications (2017). arXiv:1711.09390
41. Rainer, C.: Two different approaches to nonzero-sum stochastic differential games. Appl. Math. Optim. **56**, 131–144 (2007)
42. Schmitendorf, W.E.: Existence of optimal open-loop strategies for a class of differential games. J. Optim. Theory Appl. **5**, 363–375 (1970)
43. Sun, J.: Mean-field stochastic linear quadratic optimal control problems: open-loop solvabilities. ESAIM: COCV **23**, 1099–1127 (2017)

44. Sun, J., Li, X., Yong, J.: Open-loop and closed-loop solvabilities for stochastic linear quadratic optimal control problems. SIAM J. Control Optim. **54**, 2274–2308 (2016)
45. Sun, J., Wang, H.: Mean-field stochastic linear-quadratic optimal control problems: weak closed-loop solvability (2020). https://doi.org/10.3934/mcrf.2020026
46. Sun, J., Yong, J.: Linear quadratic stochastic differential games: open-loop and closed-loop saddle points. SIAM J. Control Optim. **52**, 4082–4121 (2014)
47. Sun, J., Yong, J.: Linear-quadratic stochastic two-person nonzero-sum differential games: open-loop and closed-loop Nash equilibria. Stoch. Proc. Appl. **129**, 381–418 (2019)
48. Sun, J., Yong, J.: Stochastic Linear-Quadratic Optimal Control Theory: Open-Loop and Closed-Loop Solutions. SpringerBriefs in Mathematics, Springer, Cham (2020)
49. Sun, J., Yong, J., Zhang, S.: Linear quadratic stochastic two-person zero-sum differential games in an infinite horizon. ESAIM: COCV **22**, 743–769 (2016)
50. Wonham, W.M.: On a matrix Riccati equation of stochastic control. SIAM J. Control **6**, 681–697 (1968)
51. Yong, J.: Linear-quadratic optimal control problems for mean-field stochastic differential equations. SIAM J. Control Optim. **51**, 2809–2838 (2013)
52. Yong, J.: Differential Games–A Concise Introduction. World Scientific Publisher, Singapore (2015)
53. Yong, J.: Linear-quadratic optimal control problems for mean-field stochastic differential equations–time-consistent solutions. Trans. Am. Math. Soc. **369**, 5467–5523 (2017)
54. Yong, J., Zhou, X.Y.: Stochastic Controls: Hamiltonian Systems and HJB Equations. Springer, New York (1999)
55. Zhang, P.: Some results on two-person zero-sum linear quadratic differential games. SIAM J. Control Optim. **43**, 2157–2165 (2005)

Index

© The Author(s), under exclusive license to Springer Nature Switzerland AG 2020
J. Sun and J. Yong, *Stochastic Linear-Quadratic Optimal Control Theory:*
Differential Games and Mean-Field Problems, SpringerBriefs in Mathematics,
https://doi.org/10.1007/978-3-030-48306-7

Printed in the United States
By Bookmasters